Lecture Notes in Mathematics

Edited by A. Dold and B. Eckmann

1287

E. B. Saff (Ed.)

Approximation Theory, Tampa

Proceedings of a Seminar held in Tampa, Florida, 1985–1986

Springer-Verlag
Berlin Heidelberg New York London Paris Tokyo

Editor

Edward B. Saff
Institute for Constructive Mathematics, Department of Mathematics
University of South Florida, Tampa, Florida 33620, USA

Mathematics Subject Classification (1980): 41A20, 41A17, 41A60, 42C05

ISBN 3-540-18500-3 Springer-Verlag Berlin Heidelberg New York
ISBN 0-387-18500-3 Springer-Verlag New York Berlin Heidelberg

Printing and binding: Druckhaus Beltz, Hemsbach/Bergstr.
2146/3140-543210

Preface

The Institute for Constructive Mathematics at the University of South Florida had its beginnings in 1985. Its goal is to foster basic research in the variety of mathematical areas that interface with approximation theory, numerical analysis, and pattern recognition. A significant component of this activity has been to provide an atmosphere conducive to research, not only for faculty at U.S.F., but also for distinguished visiting researchers from the U.S. and abroad. In its maiden year, the Institute hosted mathematicians from Canada, England, Israel, Germany, South Africa, Sweden, Switzerland, the People's Republic of China, as well as a variety of universities in the U.S. Measured by any reasonable standard, the individual and joint accomplishments of these visitors have resulted in substantial advancements in approximation theory as well as greater international cooperation and collaboration.

The papers contained in this Proceedings of the Tampa Approximation Seminar serve as a testimonial to the quality and variety of research activities conducted at the Institute. Although the main theme is approximation theory, this collection reflects the individual interests of the visitors to the Institute during the academic year 1985-1986. It is a pleasure to thank the following mathematicians for their contributions to this issue:

P.R. Graves-Morris
A.L. Levin
D. Lubinsky
H. Mhaskar
J. Nuttall
J. Palagallo
T. Price
L. Reichel
B. Shekhtman
H. Stahl
J. Waldvogel

The editor is also indebted to Prof. K. Pothoven, Chairman of the Department of Mathematics at U.S.F. for his active role in creating the Institute as well as hosting its guests. The majority of the word processing for this Proceedings is the work of Ms. Selma Canas whose careful and dedicated handling is deserving of special thanks.

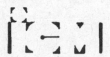

E.B. Saff
Director, Institute for
Constructive Mathematics
May 15, 1987

Preface

[illegible] modern [illegible] [illegible] [illegible] [illegible]

[illegible] conquest [illegible] [illegible]

[illegible] [illegible] a pleasure [illegible] [illegible]

[illegible]

[illegible] [illegible]

CONTENTS

A FAST ALGORITHM TO SOLVE KALMAN'S PARTIAL REALISATION PROBLEM FOR SINGLE INPUT, MULTI-OUTPUT SYSTEMS

P. R. Graves-Morris
School of Mathematical Sciences,
University of Bradford,
Bradford, West Yorkshire,
England.

and

J. M. Wilkins
D.A.M.T.P.,
Cambridge University,
Cambridge,
England.

Abstract A brief review is given of the solution of the scalar partial realisation problem using Padé approximants. The use of simultaneous Padé approximants in the solution of the single input, multi-output partial realisation problem is then discussed. We show how analogues of Frobenius identities are derived for simultaneous Padé approximants of two series, and we give twelve such identities. We show how some of these identities are combined to construct analogues of Baker's and Kronecker's algorithms. These analogues are fast algorithms for simultaneous Padé approximation of two series, and so also for a solution of the single input, two output partial realisation problem.

1. Introduction

In Kailath's book [14] and in the recent reviews by Gragg and Lindquist [10] and by Bultheel and van Barel [9], the authors explain how Padé approximants are used to construct a partial realisation of a single input, single output system. A certain number of Markov parameters of the system are specified as the data, and the Padé method is used to determine the values of the circuit elements. It is also known that simultaneous Padé approximants may be used to construct a partial realisation of a single input, multi-output system in terms of its Markov parameters [13]. This approach provides a solution of Kalman's partial realisation problem for systems [15].

For example, let h_1, h_2, h_3, ... be the Markov parameters for a single input, single output system. These parameters formally define the function

(1.1) $$h(z) := h_1 z^{-1} + h_2 z^{-2} + h_3 z^{-3} + \dots + h_{2N} z^{-2N} + \dots$$

Ignoring the exceptional, degenerate cases [2, Chap. 2; 3, Chap. 1], an [N/N] type Padé approximant of h(z) takes the form

$$(1.2)\qquad [N/N]_h(z) = \frac{a_1z^{-1} + a_2z^{-2} + \dots + a_Nz^{-N}}{1 + b_1z^{-1} + b_2z^{-2} + \dots b_Nz^{-N}}$$

with the property that

$$(1.3)\qquad [N/N]_h(z) = h(z) + O(z^{-2N-1}) .$$

Many methods exist for computing Padé approximants [2.3]. The resulting parameters $a_1, a_2, \dots, a_N, b_1, b_2, \dots, b_N$ are used to design a controller canonical realisation of the system [14], shown in Fig. 1 for the case of N = 3. The first 2N Markov parameters of this system are $h_1, h_2, \dots, h_{2N}$ and in this sense the Padé approximant provides a partial realisation of h(z).

The equivalent example of a single input, p-output system would involve $\underline{h}_1, \underline{h}_2, \dots, \underline{h}_{2N}$ as the Markov parameters to be realised, with $\underline{h}_i \in \mathbb{R}^p$. We formally define

$$(1.4)\qquad \underline{h}(z) = \underline{h}_1z^{-1} + \underline{h}_2z^{-2} + \dots + \underline{h}_{2N}z^{-2N} + \dots$$

For the case of p = 2, for example

$$(1.5)\qquad h(z) = \begin{pmatrix} h_1^{(1)} \\ h_1^{(2)} \end{pmatrix} z^{-1} + \begin{pmatrix} h_2^{(1)} \\ h_2^{(2)} \end{pmatrix} z^{-2} + \dots + \begin{pmatrix} h_{2N}^{(1)} \\ h_{2N}^{(2)} \end{pmatrix} z^{-2N} + \dots,$$

and in this case the simultaneous Padé approximant (SPA) is

$$(1.6)\qquad [N/N]_{\underline{h}}(z) =$$

$$\left(\frac{a_1^{(1)}z^{-1} + a_2^{(1)}z^{-2} + \dots + a_N^{(1)}z^{-N}}{1 + b_1z^{-1} + b_2z^{-2} + \dots + b_Nz^{-N}} , \frac{a_1^{(2)}z^{-1} + a_2^{(2)}z^{-2} + \dots + a_N^{(2)}z^{-N}}{1 + b_1z^{-1} + b_2z^{-2} + \dots + b_Nz^{-N}} \right)$$

Notice the common denominator in (1.6). The direct interpretation of the coefficients of (1.6) as values of the circuit elements is shown in Fig. 2 for the case of N = 2. In this case, the denominator polynomial is given by

$$(1.7)\qquad Q(z) = \begin{vmatrix} 1 & h_3^{(1)} & h_3^{(2)} \\ z^{-1} & h_2^{(1)} & h_2^{(2)} \\ z^{-2} & h_1^{(1)} & h_1^{(2)} \end{vmatrix}$$

up to a (normally irrelevant) constant factor. Of course, Q(z) must

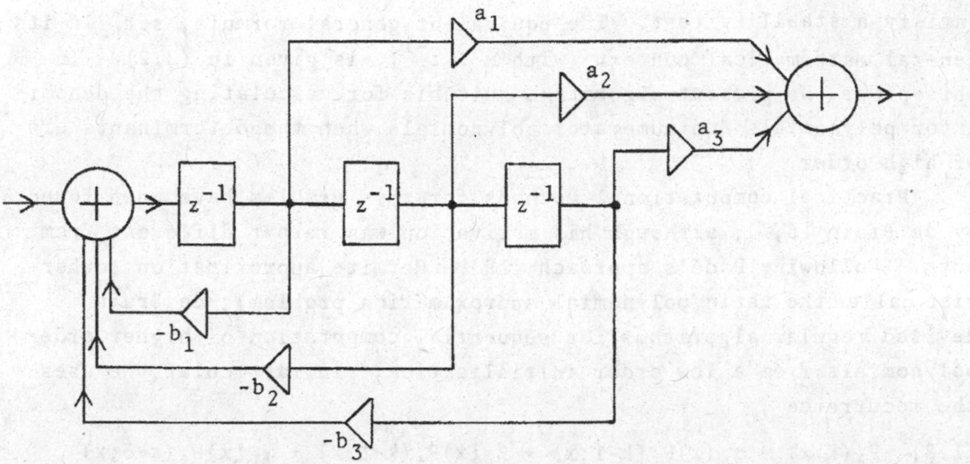

Fig. 1 Controller canonical form of a single input, single output digital system.

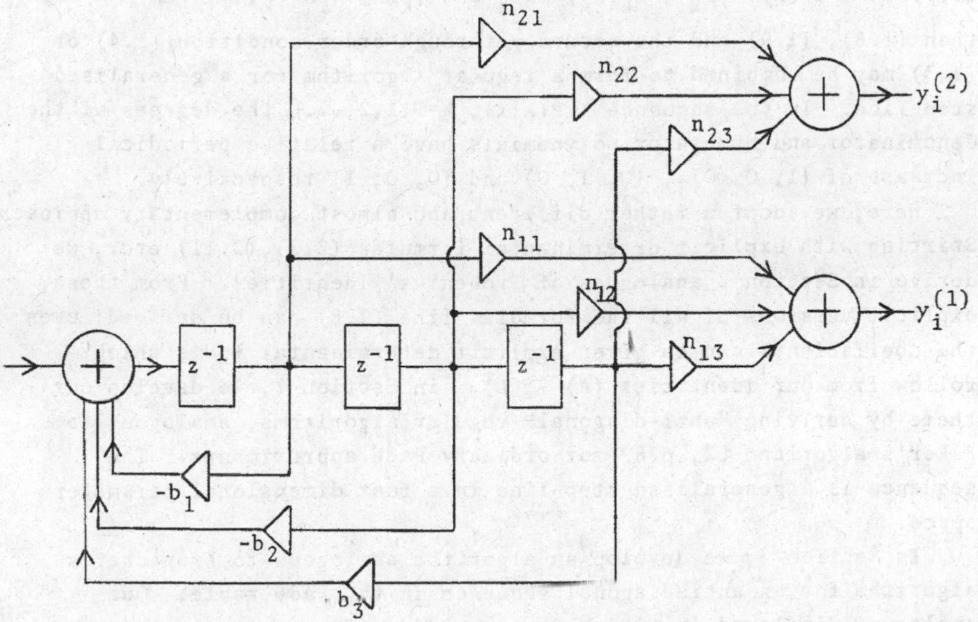

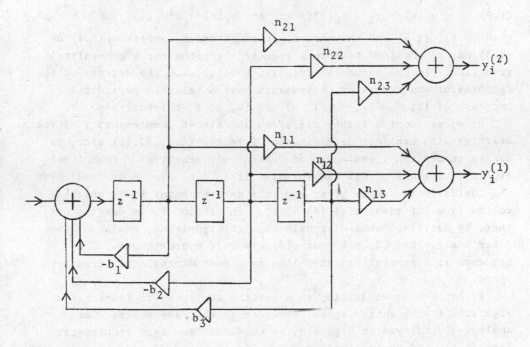

Fig. 2 Controller canonical form of a single input, two output system.

satisfy a stability test. The equivalent general formula, set in its general mathematical context (with $x \equiv z^{-1}$), is given in (2.2). In this paper, we present algorithms suitable for calculating the denominator polynomials and numerator polynomials when the determinants are of high order.

Practical computational methods for this problem have been found by de Bruin [6,7], although his motivation was rather different from ours. Following Padé's approach [19] to Hermite approximation (otherwise called the Latin polynomial approximation problem), de Bruin devised regular algorithms for sequential computation of higher order polynomials from a low order initialisation. In particular, he uses the recurrence

(1.8) $P_i(k;x) = \alpha_k(x)P_i(k-1;x) + \beta_k(x)P_i(k-2;x) + \gamma_k(x)P_i(k-3;x)$

(eq (3) of [6]) in which $P_i(k,x)$, $i = 1,2,3$, are the two numerator polynomials and the denominator polynomial constituting $\underline{P}(k;x)$, and $\underline{P}(k;x)$ is the kth vector polynomial in the sequence. In particular, if

(1.9) $\alpha_k(x) := \alpha_k$, $\beta_k(x) := x\beta_k$ and $\gamma_k(x) := x^2\gamma_k$,

then (1.8), (1.9) and the accuracy-through-order condition (1.4) or (2.3) may be combined to form a regular algorithm for a generalised step-line. In the sequence $\{\underline{P}(k;x),\ k=0,1,2,\ldots\}$, the degrees of the denominator and numerator polynomials have a relative periodical increase of (1, 0, 0) , (0, 1, 0) and (0, 0, 1) respectively.

Here, we adopt a rather different and almost complementary approach. Starting with explicit determinantal formulas (2.2),(2.11) etc., we derive in Section 2 analogues of Frobenius' identities. From these, explicit versions of all the formulas like (1.8) can be derived; even the coefficients can be given explicit determinantal forms which follow from our identities (A) - (L). In Section 2, we develop our theme by deriving "anti-diagonal" regular algorithms, analogous to Baker's algorithm [2, p78] for ordinary Padé approximants. The sequence is a generalised step-line in a four dimensional parameter space.

In Section 3, we develop an algorithm analogous to Kronecker's algorithm for an anti-diagonal sequence in the Padé table. Our analogue, displayed in Figure 6, is based on four term recurrences like (1.8), and so is also a regular algorithm for a generalised (anti-diagonal) stepline.

de Bruin's identity (1.8, 1.9) constitutes a generalisation to single input, two output systems of the Berlekamp algorithm. Other identities of de Bruin [6,7] constitute, in principle, generalisations of Berlekamp's algorithm to single input, multi-output systems. Our algorithm, in Section 3, is a generalisation of Kronecker's algorithm, as discussed by McEliece and Shearer [16]. The contrast between McEliece and Shearer's approach and Berlekamp's approach is described in [17, p369;8]. What is important for our purposes is that Kronecker's algorithm has a simple modification for reliability, as discussed by Warner [20] and McEliece and Shearer [16] and that it has a ready interpretation in terms of the block structure of the Padé table [11]. Our hope is that our algorithm of Section 3 has a simple modification for reliability similar to the Euclidean modification of Kronecker's algorithm. It may also be that de Bruin's approach has a modification for reliability similar to Massey's modification of Berlekamp's algorithm [18].

2. Analogues of Frobenius Identities

The common denominator polynomial for the SPA of type $[N_1, N_2; N_1-m_1, N_2-m_2; m]$ for two series

(2.1) $\quad f^{(1)}(x) := \Sigma_{i=0}^{\infty} c_i^{(1)} x^i \quad , \quad f^{(2)}(x) := \Sigma_{i=0}^{\infty} c_i^{(2)} x^i$

is

(2.2) $\quad Q^{[N_1,\, N_2;\, N_1-m_1,\, N_2-m_2;\, m]}(x)$

$$= \begin{vmatrix} 1 & c^{(1)}_{N_1} & c^{(1)}_{N_1-1} & \cdots & c^{(1)}_{N_1-m_1+1} & c^{(2)}_{N_2} & c^{(2)}_{N_2-1} & \cdots & c^{(2)}_{N_2-m_2+1} \\ x & c^{(1)}_{N_1-1} & c^{(1)}_{N_1-2} & \cdots & c^{(1)}_{N_1-m_1} & c^{(2)}_{N_2-1} & c^{(2)}_{N_2-2} & \cdots & c^{(2)}_{N_2-m_2} \\ \cdot & \cdot & \cdot & \cdot & & \cdot & \cdot & \cdot & \\ \cdot & \cdot & \cdot & \cdot & & \cdot & \cdot & \cdot & \\ \cdot & \cdot & \cdot & \cdot & & \cdot & \cdot & \cdot & \\ x^m & c^{(1)}_{N_1-m} & c^{(1)}_{N_1-m-1} & \cdots & c^{(1)}_{N_1-m_1-m+1} & c^{(2)}_{N_2-m} & c^{(2)}_{N_2-m-1} & \cdots & c^{(2)}_{N_2-m_2-m+1} \end{vmatrix}$$

up to an arbitrary constant factor, and with the understanding that $c_j^{(1)} = c_j^{(2)} := 0$ for $j < 0$. A result, very similar to (2.2), was originally derived by de Bruin [5]. From (2.2), it is easily verified that polynomials $P^{(1)}(x)$ and $P^{(2)}(x)$ exist for which

$$(2.3a)\qquad Q^{[\cdot\cdot]}(x)f^{(1)}(x) = P^{(1)}(x) + O\left(x^{N_1+1}\right)$$

$$(2.3b)\qquad Q^{[\cdot\cdot]}(x)f^{(2)}(x) = P^{(2)}(x) + O\left(x^{N_2+1}\right)$$

with

$$(2.4)\qquad \partial P^{(1)} \le N_1-m_1\ ,\ \partial P^{(2)} \le N_2-m_2\ ,$$

$$(2.5)\qquad \partial Q^{[\cdot\cdot]} \le m\ ,$$

and we have used the abbreviation

$$Q^{[\cdot\cdot]}(x):= Q^{[N_1\ N_2;N_1-m_1,N_2-m_2;m]}(x).$$

We view (2.3) as defining $P^{(1)}(x)$, $P^{(2)}(x)$. Formulas (2.2) and (2.3) can be combined to construct determinantal representations of $P^{(1)}(x)$ and $P^{(2)}(x)$. We also define the vector polynomial $\underline{P}^{[\cdot\cdot]}(x) := (P^{(1)}(x), P^{(2)}(x))$. If $Q^{[\cdot\cdot]}(0) \ne 0$, and $N_1 = N_2 = N$,

$$(2.6)\qquad \underline{P}^{[\cdot\cdot]}(x)/Q(x) = \underline{f}(x) + O(x^{N+1}).$$

in which the left-hand side is a vector-valued Padé approximant. With the aid of (2.2) and its companion formulas for $P^{(1)}(x)$ and $P^{(2)}(x)$, we can now derive the analogues of Frobenius' identities for SPAs.

Jacobi's identity [1, p 99] is commonly written as

$$(2.7)\qquad D_{pq;rs}\ D = D_{p;r}\ D_{q;s} - D_{p;s}\ D_{q;r}\ .$$

The notation $D_{p;r}$ in (2.7) denotes the determinant of some original square matrix, of which the pth row and r^{th} column have been deleted; D and $D_{pq;rs}$ have analogous meanings. By applying (2.7) to (2.2) with $(p,q,r,s) = (1,m+1,1,m+1)$, we obtain identity (A):

$$\begin{aligned}(2.8)\qquad & Q^{[N_1,N_2;\ N_1-m_1,N_2-m_2;\ m]}(x)\ C(N_1,N_2;\ N_1-m_1,N_2-m_2+1;\ m-1)\\ &= Q^{[N_1,N_2;\ N_1-m_1,N_2-m_2+1;\ m-1]}(x)\ C(N_1,N_2;\ N_1-m_1,N_2-m_2;m)\\ &\quad -xQ^{[N_1-1,N_2-1;\ N_1-m_1-1,N_2-m_2;m-1]}(x)\ C(N_1+1,N_2+1,N_1-m_1+1,N_2-m_2+1;m)\end{aligned}$$

where

$$(2.9)\qquad C(N_1,N_2;\ L_1,L_2;\ m) := Q^{[N_1,N_2;\ L_1,L_2;\ m]}(0).$$

With $(p,q,r,s) = (1,m+1,1,m_1+2)$ in (2.7), we obtain identity (B):

$$(2.10)\quad Q^{[N_1,N_2;\,N_1-m_1,N_2-m_2;m]}(x)\;C(N_1,N_2-1;\,N_1-m_1,N_2-m_2;\,m-1)$$
$$=\;Q^{[N_1,N_2-1;\,N_1-m_1,N_2-m_2;m-1]}(x)\;C(N_1,N_2;\,N_1-m_1,N_2-m_2;\,m)$$
$$-xQ^{[N_1-1,N_2-2;\,N_1-m_1-1,N_2-m_2-1;\,m-1]}(x)\;C(N_1+1,N_2+1;\,N_1-m_1+1,N_2-m_2+1;\,m)$$

We regard identities obtained by interchange of the given series $f^{(1)} \leftrightarrow f^{(2)}$ as inessentially different. With this proviso, (2.8) and (2.10) are the only two such identities obtainable from (2.2) using (2.7) directly. However,

$$(2.11)\quad Q^{[N_1,N_2;\,N_1-m_1,N_2-m_2;\,m]}(x)$$
$$= -\begin{vmatrix} 0 & 1 & 0 & \cdots & 0 & 0 & 0 & \cdots & 0 \\ 1 & c^{(1)}_{N_1+1} & c^{(1)}_{N_1} & \cdots & c^{(1)}_{N_1-m_1+1} & c^{(2)}_{N_2} & c^{(2)}_{N_2-1} & \cdots & c^{(2)}_{N_2-m_2+1} \\ x & c^{(1)}_{N_1} & c^{(1)}_{N_1-1} & \cdots & c^{(1)}_{N_1-m} & c^{(2)}_{N_2-1} & c^{(2)}_{N_2-2} & \cdots & c^{(2)}_{N_2-m_2} \\ \cdot & & \cdot & \cdot & \cdot & & \cdot & \cdot & \cdot \\ \cdot & & \cdot & \cdot & \cdot & & \cdot & \cdot & \cdot \\ \cdot & & \cdot & \cdot & \cdot & & \cdot & \cdot & \cdot \\ x^m & c^{(1)}_{N_1-m+1} & c^{(1)}_{N_1-m} & \cdots & c^{(1)}_{N_1-m_1-m+1} & c^{(2)}_{N_2-m} & c^{(2)}_{N_2-m-1} & \cdots & c^{(2)}_{N_2-m_2-m+1} \end{vmatrix}$$

Schwein's identity [1, p 108] follows from Jacobi's identity in the form (2.7) and applied to a determinant such as (2.11) [3, p 85].

With (p, q, r, s) = (1, m+2, 1, m_1+2), we obtain identity (C):

$$(2.12)\quad Q^{[N_1\;N_2;\,N_1-m_1,N_2-m_2;\,m]}(x)\;C(N_1+2,N_2+1;\,N_1-m_1+2,N_2-m_2+1;\,m)$$
$$=Q^{[N_1,N_2;\,N_1-m_1+1,N_2-m_2;\,m-1]}(x)\;C(N_1+2,N_2+1;\,N_1-m_1+1,N_2-m_2+1;m+1)$$
$$+\;Q^{[N_1+1,N_2;\,N_1-m_1+1,N_2-m_2;\,m]}(x)\;C(N_1+1,N_2+1;\,N_1-m_1+1,N_2-m_2+1;\,m)$$

With (p,q,r,s) = (1, m+2, 1, m_1+3), we obtain identity (D):

$$(2.13)\quad Q^{[N_1,N_2;\,N_1-m_1,N_2-m_2;\,m]}(x)\;C(N_1+2,N_2;\,N_1-m_1+1,N_2-m_2+1;\,m)$$
$$=Q^{[N_1,N_2-1;\,N_1-m_1,N_2-m_2;\,m-1]}(x)\;C(N_1+2,N_2+1;\,N_1-m_1+1,N_2-m_2+1;\,m+1)$$
$$+\;Q^{[N_1+1,N_2-1;\,N_1-m_1,N_2-m_2;\,m]}(x)\;C(N_1+1,N_2+1;\,N_1-m_1+1,N_2-m_2+1;\,m)$$

With (p,q,r,s) = (1, m+2, 1, m+2), we obtain identity (E):

$$(2.14)\quad Q^{[N_1,N_2;\ N_1-m_1,N_2-m_2;\ m]}(x)\ C(N_1+2,N_2+1;\ N_1-m_1+1,N_2-m_2+2;\ m)$$
$$= Q^{[N_1,N_2;\ N_1-m_1,N_2-m_2+1;\ m-1]}(x)\ C(N_1+2,N_2+1;\ N_1-m_1+1,N_2-m_2+1;\ m+1)$$
$$+ Q^{[N_1+1,N_2;\ N_1-m_1,N_2-m_2+1;\ m]}(x)\ C(N_1+1,N_2+1;\ N_1-m_1+1,N_2-m_2+1;\ m)$$

With $(p,q,r,s) = (1, 2, 1, m_1+2)$, we obtain identity (F):

$$(2.15)\quad Q^{[N_1,N_2;\ N_1-m_1,N_2-m_2;\ m]}(x)\ C(N_1+1,N_2;\ N_1-m_1+1,N_2-m_2;\ m)$$
$$= xQ^{[N_1-1,N_2-1;\ N_1-m_1,N_2-m_2-1;\ m-1]}(x)\ C(N_1+2,N_2+1;\ N_1-m_1+1,N_2-m_2+1;\ m+1)$$
$$+\ Q^{[N_1+1,N_2;\ N_1-m_1+1,N_2-m_2;\ m]}(x)\ C(N_1,N_2;\ N_1-m_1,N_2-m_2;\ m)$$

With $(p,q,r,s) = (1, 2, 1, m_1+3)$, we obtain identity (G):

$$(2.16)\quad Q^{[N_1,N_2;\ N_1-m_1;\ N_2-m_2;\ m]}(x)\ C(N_1+1,N_2-1;\ N_1-m_1,N_2-m_2;\ m)$$
$$= xQ^{[N_1-1,N_2-2;\ N_1-m_1-1,N_2-m_2-1;\ m-1]}(x)\ C(N_1+2,N_2+1;\ N_1-m_1+1,\ N_2-m_2+1;\ m+1)$$
$$+\ Q^{[N_1+1,N_2-1;\ N_1-m_1,N_2-m_2;\ m]}(x)\ C(N_1,N_2;\ N_1-m_1,N_2-m_2;\ m)$$

With $(p,q,r,s) = (1, 2, 1, m+2)$, we obtain identity (H):

$$(2.17)\quad Q^{[N_1,N_2;\ N_1-m_1,N_2-m_2;\ m]}(x)\ C(N_1+1,N_2;\ N_1-m_1,N_2-m_2+1;\ m)$$
$$= xQ^{[N_1-1,N_2-1;\ N_1-m_1-1,N_2-m_2;\ m-1]}(x)\ C(N_1+2,N_2+1;\ N_1-m_1+1,N_2-m_2+1;m+1)$$
$$+\ Q^{[N_1+1,N_2;\ N_1-m_1,N_2-m_2+1;\ m]}(x)\ C(N_1,N_2;\ N_1-m_1,N_2-m_2;\ m)$$

Next, we consider the determinant in

$$(2.18)\quad Q^{[N_1,N_2;\ N_1-m_1,N_2-m_2;\ m]}(x)\ .(-1)^{m_1+1}$$
$$= \begin{vmatrix} 0 & 0 & \cdots & 0 & 1 & 0 & \cdots & 0 \\ 1 & c^{(1)}_{N_1} & \cdots & c^{(1)}_{N_1-m_1+1} & c^{(1)}_{N_1-m_1} & c^{(2)}_{N_2} & \cdots & c^{(2)}_{N_2-m_2+1} \\ x & c^{(1)}_{N_1-1} & \cdots & c^{(1)}_{N_1-m_1} & c^{(1)}_{N_1-m_1-1} & c^{(2)}_{N_2-1} & \cdots & c^{(2)}_{N_2-m_2} \\ \cdot & & \cdots & & & & \cdots & \\ \cdot & & \cdots & & & & \cdots & \\ x^m & c^{(1)}_{N_1-m} & \cdots & c^{(1)}_{N_1-m_1-m+1} & c^{(1)}_{N_1-m_1-m} & c^{(2)}_{N_2-m} & \cdots & c^{(2)}_{N_2-m-m_2+1} \end{vmatrix}$$

With $(p,q,r,s) = (1, 2, 1, m+2)$ in (2.7), (2.18), we obtain identity (I):

$$(2.19)\qquad Q^{[N_1,N_2;N_1-m_1,N_2-m_2;m]}(x)\quad C(N_1,N_2;N_1-m_1-1,N_2-m_2+1;m)$$

$$= xQ^{[N_1-1,N_2-1;N_2-m_1-1,N_2-m_2;m-1]}(x)\quad C(N_1+1,N_2+1;N_1-m_1,N_2-m_2+1;m+1)$$

$$+\ Q^{[N_1,N_2;N_1-m_1-1,N_2-m_2+1;m]}(x)\quad C(N_1,N_2;N_1-m_1,N_2-m_2,m)$$

and with (p,q,r,s) = (1, m+2, 1, m+2) in (2.7), (2.18), we obtain identity (J):

$$(2.20)\qquad Q^{[N_1,N_2;N_1-m_1,N_2-m_2;m]}(x)\quad C(N_1+1,N_2+1;N_1-m_1,N_2-m_2+2;m)$$

$$= Q^{[N_1,N_2;N_1-m_1,N_2-m_2+1;m-1]}(x)\quad C(N_1+1,N_2+1;N_1-m_1,N_2-m_2+1;m+1)$$

$$+\ Q^{[N_1,N_2;N_1-m_1-1,N_2-m_2+1;m]}(x)\quad C(N_1+1,N_2+1;N_1-m_1+1,N_2-m_2+1;m).$$

There are many other possible values for (p,q,r,s) which yield Frobenius type identities from (2.7), (2.18), but each that we found is a duplicate of one of the preceding identities. However, more identities similar to (A)-(J) follow by elimination of common elements, as happens in the one-dimensional (Padé) case.

From identities (E) and (J), we eliminate $Q^{[N_1,N_2;N_1-m_1,N_2-m_2+1;m-1]}(x)$ and this leads to identity (K):

$$(2.21)\qquad Q^{[N_1,N_2;N_1-m_1,N_2-m_2;m]}(x)\quad C(N_1+2,N_2+1;N_1-m_1,N_2-m_2+2;m+1)$$

$$= Q^{[N_1,N_2;N_1-m_1-1,N_2-m_2+1;m]}(x)\quad C(N_1+2,N_2+1;N_1-m_1+1,N_2-m_2+1;m+1)$$

$$+\ Q^{[N_1+1,N_2;N_1-m_1,N_2-m_2+1;m]}(x)\quad C(N_1+1,N_2+1;N_1-m_1,N_2-m_2+1;m+1)$$

Similarly, we put N_2-1 instead of N_2 in (2.14), identity (E), and eliminate $Q^{[N_1,N_2-1;N_1-m_1,N_2-m_2;m-1]}(x)$ from (E) and (D) to obtain identity (L):

$$(2.22)\qquad Q^{[N_1,N_2;N_1-m_1,N_2-m_2;m]}(x)\quad C(N_1+2,N_2;N_1-m_1+1,N_2-m_2;m+1)$$

$$= Q^{[N_1,N_2-1;N_1-m_1,N_2-m_2-1;m]}(x)\quad C(N_1+2,N_2+1;N_1-m_1+1,N_2-m_2+1;m+1)$$

$$+\ Q^{[N_1+1,N_2-1;N_1-m_1,N_2-m_2;m]}(x)\quad C(N_1+1,N_2+1;N_1-m_1+1,N_2-m_2,m+1)$$

It may be possible, using the techniques of Padé [19] and de Bruin [5, 6, 7 and references in 7] to show that (A)-(L) are the only Frobenius-type identities for SPAs of two series, but we have not done this here. Just as Frobenius identities are conveniently displayed by diagrams such as $\begin{pmatrix} * & * \\ * & \end{pmatrix}$, so also the identities (A)-(L) should be displayed in a four-dimensional figure. We show the results in two-dimensional sections of figure 4. As an example, identity (A) connects the elements shown in figure 3.

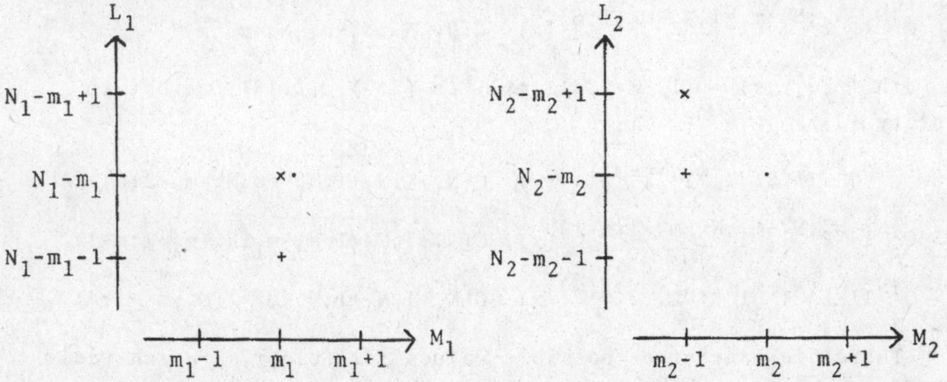

Fig. 3 The locations of the elements of identity (A), denoted by ×, . and +, in the (L_1, M_1) and (L_2, M_2) planes.

The co-ordinates (L_1,M_1), (L_2,M_2) shown in Fig.3 are common to all the identities, and so they are omitted from Fig.4 for conciseness. Fig.4 shows the relative locations of the polynomials in the Frobenius-type identities.

We have expressed the twelve Frobenius type identities among (2.8)-(2.22) in terms of denominator polynomials. It is well known that they also apply to the numerator polynomial $P_1(x)$ (from (2.3a), by truncation) and similarly to $P_2(x)$. Partly to emphasise this point, we define

$$(2.23)\qquad \underline{S}^{[N_1,N_2;L_1,L_2;m]}(x) := \begin{bmatrix} P_1^{[N_1,N_2;L_1,L_2;m]}(x) \\ P_2^{[N_1,N_2;L_1,L_2;m]}(x) \\ Q^{[N_1,N_2;L_1,L_2;m]}(x) \end{bmatrix} .$$

Identity (A) then becomes

$$(2.24)\qquad \underline{S}^{[N_1,N_2;N_1-m_1,N_2-m_2;m]}(x)$$
$$= \alpha\underline{S}^{[N_1,N_2;N_1-m_1,N_2-m_2+1;m-1]}(x)$$
$$+ \beta x\underline{S}^{[N_1-1,N_2-1;N_1-m_1-1,N_2-m_2;m-1]}(x)$$

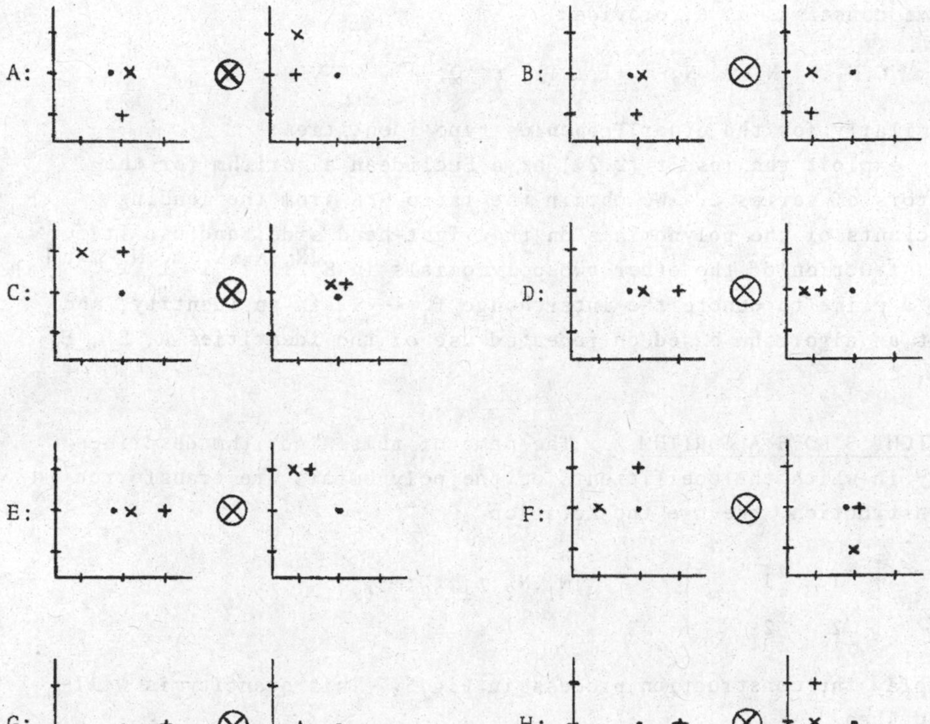

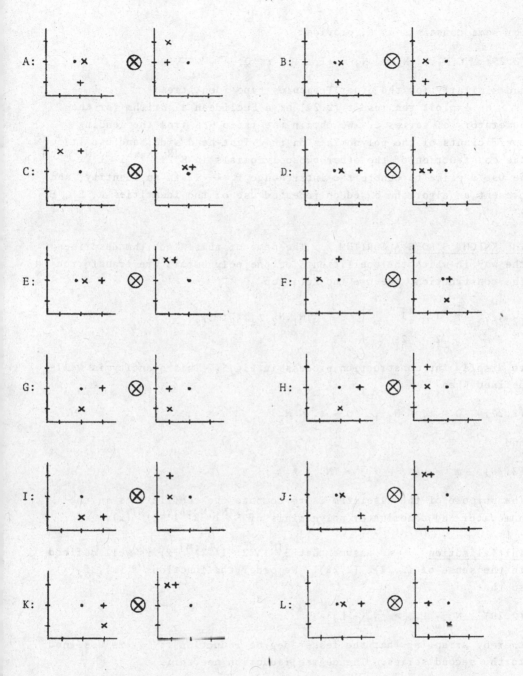

Fig. 4 The relative locations of the polynomials connected by the twelve Frobenius-type identities. Detail of identity A is shown in Fig.3.

for some constants α, β, provided

(2.25) $C(N_1,N_2;N_1-m_1,N_2-m_2+1;m-1) \neq 0,$

and similarly for the other Frobenius type identities.

We exploit the result (2.24) as a Euclidean algorithm for the numerators of series 2. We obtain the ratio $\alpha:\beta$ from the leading coefficients of the polynomials on the right-hand side, and use it for the construction of the other two polynomials in $\underline{S}^{[N_1,N_2;N_1-m_1,N_2-m_2;m]}(x)$. We use a prime to denote the interchange $f_1 \leftrightarrow f_2$ in an identity, and present an algorithm based on repeated use of the identities A, E', E; A', C', C.

THE KNIGHT'S MOVE ALGORITHM The name of this algorithm describes the way in which the coefficients of the polynomials are transferred in the construction. We use the notation

(2.26) $$\begin{bmatrix} N_1 & \ell_1 & \\ & & m \\ N_2 & \ell_2 & \end{bmatrix} := S^{[N_1,N_2;\ell_1,\ell_2;m]}(x)$$

to display the construction process in Fig.5. This quantity is well-defined when

(2.27) $0 \leq \ell_1 \leq N_1$, $0 \leq \ell_2 \leq N_2$

and

(2.28) $m = (N_1 - \ell_1) + (N_2 - \ell_2).$

The purpose of the algorithm is to compute the coefficients in the numerator and denominator polynomials of $S^{[N_1,N_2;L_1,L_2;M]}(x)$.

Initialisation We assume that $S^{[N_1,N_2;L_1,L_2;M]}(x)$ is well defined in the sense of (2.27), (2.28). We order the functions $f_1(x)$, $f_2(x)$ so that

(2.29) $N_1 - L_1 \geq N_2 - L_2$

thereby arranging that the lesser degree reduction (by μ) is assigned to the second series. The degree reduction needed is

(2.30) $\mu := N_2 - L_2$.

We also define

(2.31) $M_1 := (N_1 - L_1) - (N_2 - L_2)$

and note that $N_1 \geq M_1 \geq 0$, by (2.27) and (2.29). The numerator and denominator polynomials of the Padé approximants of types $[N_1 - M_1/M_1]$ and $[N_1 - M_1 - 1/M_1 + 1]$ for series 1 are needed to initialise the algorithm. These, together with the Maclaurin polynomials of degrees N_2, $N_2 - 1$ for series 2, constitute the data entries in the first two rows of figure 5.

Iteration   Alternate cycles, based on (A E' E), (A' C' C), are used, as shown in Fig.5. Each entry, defined by (2.26), represents a triple of polynomial coefficients. The entries in the right-hand columns have already been calculated in previous rows, as indicated by the arrows from a Knight's Move location, or else at the initialisation stage. In each row, the left-hand entry is obtained by a Euclidean reduction on the appropriate numerator polynomials of the two right-hand entries. (This Euclidean reduction wasdescribed following (2.24) in the context of identity A.) We assume that all the relevant numerator polynomials are of full indicated degree, so that the constants equivalent to α, β in (2.24) exist at every stage.

Termination   The algorithm terminates when the SPA of type

$$\begin{pmatrix} N_1 & N_1 - M_1 - \mu & \\ & & M_1 + 2\mu \\ N_2 & N_2 - \mu & \end{pmatrix} \equiv S^{[N_1,N_2;L_1,L_2;M]}$$

has been constructed. This occurs in row $6\mu - 2$ of Fig.5.

By inspection of the columns of the determinant in (2.2), it is obvious that a non-degenerate SPA cannot exist unless

(2.32) $m_1 \leq N_1, \quad m_2 \leq N_2$.

By inspection of the final row of the determinant, we also need

(2.33) $m \leq \max(N_1, N_2)$

for existence of a non-degenerate SPA of type $[N_1,N_2;N_1-m_1,N_2-m_2;m]$. The inequalities (2.32), (2.33) limit the amount of degree reduction normally attainable.

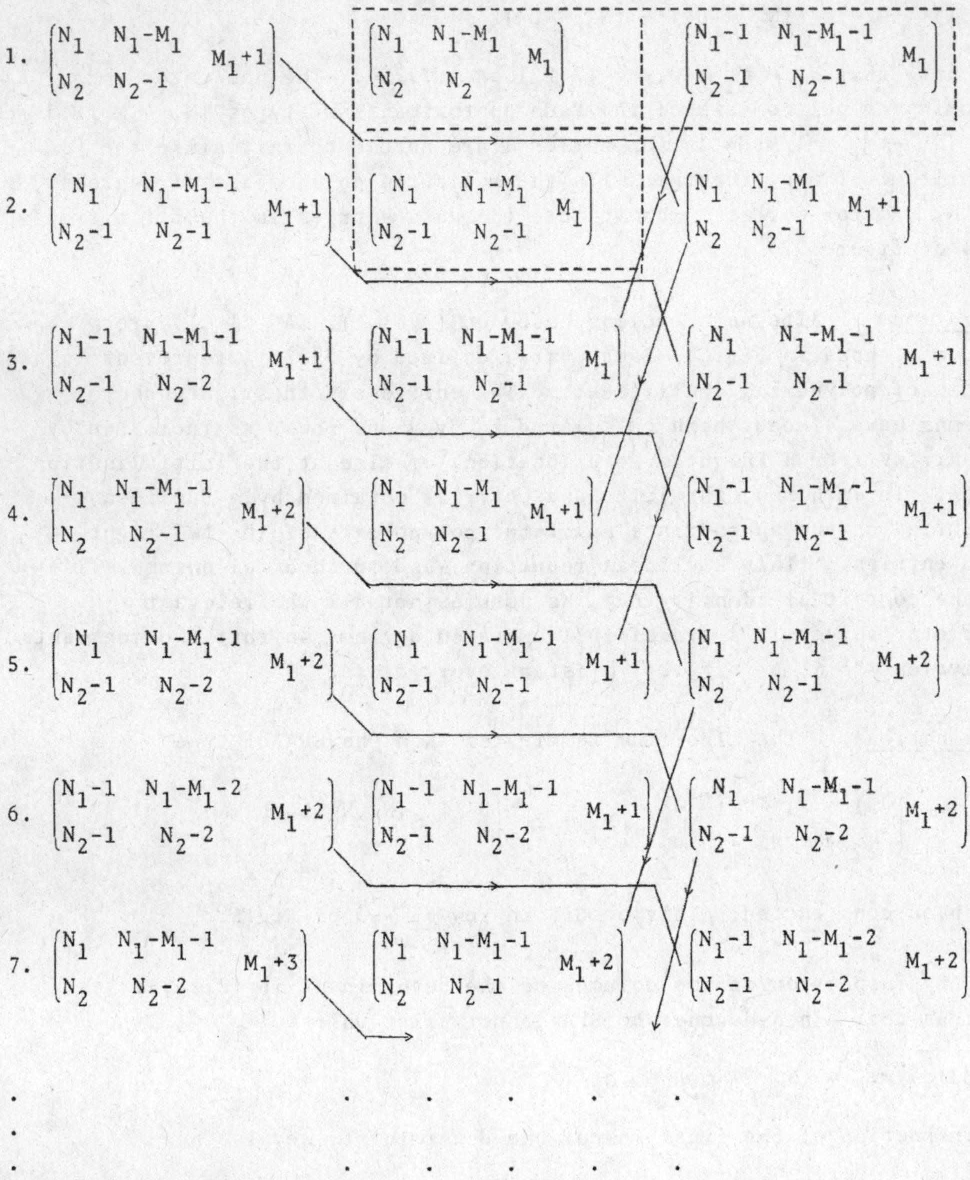

Fig. 5 Data flow for the Knight's Move algorithm. The data entries are shown in boxes in rows 1 and 2. The flow of data is shown by the arrows.

3. Kronecker-type algorithms

For ordinary, scalar Padé approximation, Kronecker's algorithm is a three term recurrence relation, connecting elements in an anti-diagonal sequence. Kronecker's algorithm has become important because a Euclidean modification converts it into a reliable algorithm, with a clear interpretation in terms of the block structure of the Padé table [11, 16, 20].

We present two analogues of Kronecker's algorithm for two series here. The first is obtained by elimination of some intermediate elements explicitly constructed in the Knight's Move algorithm. Our aim is the construction of a SPA of type $[N_1,N_2;L_1,L_2;M]$, where $M = N_1-L_1+N_2-L_2$, from a sequence in which both N_1 and N_2 have fixed values. Elements of type $[N_1, N_2-1; \ldots\ldots]$ are not of the required type. They may be eliminated between identities (C, C') to yield an identity (X') and between (E, E') to yield an identity (X). The resulting identities are

$$\text{(3.1)} \quad \begin{bmatrix} N_1 & N_1-m_1 & \\ N_2 & N_2-m_2+1 & m-1 \end{bmatrix} = \alpha'_A \begin{bmatrix} N_1 & N_1-m_1+1 & \\ N_2 & N_2-m_2+1 & m-2 \end{bmatrix} \quad \text{(A')}$$

$$+ \beta'_A x \begin{bmatrix} N_1-1 & N_1-m_1 & \\ N_2-1 & N_2-m_2 & m-2 \end{bmatrix}$$

$$\text{(3.2)} \quad \begin{bmatrix} N_1-1 & N_1-m_1-1 & \\ N_2-1 & N_2-m_2 & m-1 \end{bmatrix} = \alpha'_X \begin{bmatrix} N_1-1 & N_1-m_1 & \\ N_2-1 & N_2-m_2 & m-2 \end{bmatrix} \quad \text{(X')}$$

$$+ \beta'_X \begin{bmatrix} N_1 & N_1-m_1 & \\ N_2 & N_2-m_2+1 & m-1 \end{bmatrix} + \gamma'_X \begin{bmatrix} N_1-1 & N_1-m_1 & \\ N_2-1 & N_2-m_2+1 & m-3 \end{bmatrix}$$

$$\text{(3.3)} \quad \begin{bmatrix} N_1 & N_1-m_1 & \\ N_2 & N_2-m_2 & m \end{bmatrix} = \alpha_A \begin{bmatrix} N_1 & N_1-m_1 & \\ N_2 & N_2-m_2+1 & m-1 \end{bmatrix} \quad \text{(A)}$$

$$+ \beta_A x \begin{bmatrix} N_1-1 & N_1-m_1-1 & \\ N_2-1 & N_2-m_2 & m-1 \end{bmatrix}$$

$$\text{(3.4)} \quad \begin{bmatrix} N_1-1 & N_1-m_1\ 1 & \\ N_2-1 & N_2-m_2-1 & m \end{bmatrix} = \alpha_X \begin{bmatrix} N_1-1 & N_1-m_1-1 & \\ N_2-1 & N_2-m_2 & m-1 \end{bmatrix} \quad \text{(X)}$$

$$+ \beta_X \begin{bmatrix} N_1 & N_1-m_1 & \\ N_2 & N_2-m_2 & m \end{bmatrix} + \gamma_X \begin{bmatrix} N_1-1 & N_1-m_1 & \\ N_2-1 & N_2-m_2 & m-2 \end{bmatrix}$$

We have used the notation of Fig.3 to represent the triples of polynomials in (3.1)-(3.4). Identities (A', A) are used, as they stand, in rows 1, 4 mod (6) of Fig.3. By using the contracted identities (X') and (X) above, (3.1)-(3.4) constitutes (in fact) a self-contained algorithm in the Euclidean sense. However, elements of type $[N_1-1,\ N_2-1;\ .....]$ are not of the type required. They are eliminated from (A'X'A) and (AXA') yielding identities (Y'), (Y) respectively, which are

$$(3.5)\quad \begin{bmatrix} N_1 & N_1-m_1 & \\ N_2 & N_2-m_2 & m \end{bmatrix} = (\alpha'_Y + x\beta'_Y)\begin{bmatrix} N_1 & N_1-m_1 & \\ N_2 & N_2-m_2+1 & m-1 \end{bmatrix} \qquad (Y')$$

$$+ \gamma'_Y\begin{bmatrix} N_1 & N_1-m_1+1 & \\ N_2 & N_2-m_2+1 & m-2 \end{bmatrix} + \delta'_Y\begin{bmatrix} N_1 & N_1-m_1+1 & \\ N_2 & N_2-m_2+2 & m-3 \end{bmatrix}$$

$$(3.6)\quad \begin{bmatrix} N_1 & N_1-m_1-1 & \\ N_2 & N_2-m_2 & m+1 \end{bmatrix} = (\alpha_Y + x\beta_Y)\begin{bmatrix} N_1 & N_1-m_1 & \\ N_2 & N_2-m_2 & m \end{bmatrix} \qquad (Y)$$

$$+ \gamma_Y\begin{bmatrix} N_1 & N_1-m_1 & \\ N_2 & N_2-m_2+1 & m+1 \end{bmatrix} + \delta_Y\begin{bmatrix} N_1 & N_1-m_1+1 & \\ N_2 & N_2-m_2+1 & m-2 \end{bmatrix}$$

All the elements involved in (Y, Y') are of type $[N_1, N_2; .....]$, as required in the solution. The identity (Y') has a Euclidean interpretation: the right-hand side of (3.5) has numerator polynomials ostensibly of degrees N_1-m_1+1 and N_2-m_2+2, which must reduce by 1 and 2 respectively to fit the left-hand side. We therefore have three implicit simultaneous equations for $\alpha'_Y:\beta'_Y:\gamma'_Y:\delta'_Y$. All this, of course, presupposes normality of the numerator polynomials. Having derived values for α'_Y, β'_Y, γ'_Y, δ'_Y, the denominator polynomial and numerator polynomials on the left-hand side of (3.5) are constructed. Identity (3.6) has a similar interpretation. The identities (Y, Y') form the basis of a self-contained algorithm:

<u>THE FIRST KRONECKER TYPE ALGORITHM</u> The preceding remarks justify the abbreviated notation

$$(3.7)\quad \begin{bmatrix} L_1 & \\ L_2 & m \end{bmatrix} := \underline{S}^{[N_1N_2;L_1L_2;m]}(x)\ .$$

Provided $N_1 \geqslant L_1$, $N_2 \geqslant L_2$ and $M = N_1 - L_1 + N_2 - L_2$, we construct

$\left[\begin{matrix} L_1 & \\ & M \\ L_2 & \end{matrix}\right]$ as follows:

<u>Initialisation</u> Order the functions $f^{(1)}$, $f^{(2)}$ so that $N_1-L_1 \geqslant N_2-L_2$, let $M_1 := (N_1-L_1) - (N_2-L_2)$, $\mu := N_2-L_2$, and complete the same initialisation as for the Knight's Move algorithm. The artificial entry, needed in the first row of Fig.6, is taken as

$$\left[\begin{matrix} N_1-M_1 & \\ & M_1-1 \\ N_2+1 & \end{matrix}\right] := \left[\begin{matrix} 0 \\ z^{N_2+1} \\ 0 \end{matrix}\right] \tag{3.8}$$

<u>Iteration</u> A Euclidean reduction of the numerators in the three right-hand columns of Fig.6 is performed to yield the next numerators. Again, we assume that the relevant numerator polynomials have full indicated degrees. The denominators follow with the same coefficients as previously described.

$$1.\quad \left[\begin{matrix} N_1-M_1-1 & \\ & M_1+2 \\ N_2-1 & \end{matrix}\right] = (a_1x+b_1)\left[\begin{matrix} N_1-M_1-1 & \\ & M_1+1 \\ N_2 & \end{matrix}\right] + c_1\left[\begin{matrix} N_1-M_1 & \\ & M_1 \\ N_2 & \end{matrix}\right] + d_1\left[\begin{matrix} N_1-M_1 & \\ & M_1-1 \\ N_2+1 & \end{matrix}\right]$$

$$2.\quad \left[\begin{matrix} N_1-M_1-2 & \\ & M_1+3 \\ N_2-1 & \end{matrix}\right] = (a_2x+b_2)\left[\begin{matrix} N_1-M_1-1 & \\ & M_1+2 \\ N_2-1 & \end{matrix}\right] + c_2\left[\begin{matrix} N_1-M_1-1 & \\ & M_1+1 \\ N_2 & \end{matrix}\right] + d_2\left[\begin{matrix} N_1-M_1 & \\ & M_1 \\ N_2 & \end{matrix}\right]$$

$$3.\quad \left[\begin{matrix} N_1-M_1-2 & \\ & M_1+4 \\ N_2-2 & \end{matrix}\right] = (a_3x+b_3)\left[\begin{matrix} N_1-M_1-2 & \\ & M_1+3 \\ N_2-1 & \end{matrix}\right] + c_3\left[\begin{matrix} N_1-M_1-1 & \\ & M_1+2 \\ N_2-1 & \end{matrix}\right] + d_3\left[\begin{matrix} N_1-M_1-1 & \\ & M_1+1 \\ N_2 & \end{matrix}\right]$$

$$4.\quad \left[\begin{matrix} N_1-M_1-3 & \\ & M_1+5 \\ N_2-2 & \end{matrix}\right] = (a_4x+b_4)\left[\begin{matrix} N_1-M_1-2 & \\ & M_1+4 \\ N_2-2 & \end{matrix}\right] + c_4\left[\begin{matrix} N_1-M_1-2 & \\ & M_1+3 \\ N_2-1 & \end{matrix}\right] + d_4\left[\begin{matrix} N_1-M_1-1 & \\ & M_1+2 \\ N_2-1 & \end{matrix}\right]$$

$$\vdots \qquad\qquad \vdots\ \vdots\ \vdots$$

<u>Fig. 6</u> The elements in the left-hand column are calculated from those in the three right-hand columns. Flow of data is shown by arrows.

<u>Termination</u> The algorithm is terminated when the desired SPA of the type

$$\begin{pmatrix} N_1-M_1-\mu & M_1+2\mu \\ N_2-\mu & \end{pmatrix} \equiv S^{[N_1,N_2;L_1,L_2;M]}(x)$$

is constructed. This occurs in row $2\mu-1$.

This algorithm has the same pattern of data flow as Kronecker's algorithm for the scalar case. This pattern is displayed in Fig.6. However, the identities (Y, Y') are four term recurrence relations rather than the three term ones for the scalar case.

THE SECOND KRONECKER TYPE ALGORITHM Our second algorithm for construction of a general SPA of type $[N_1,N_2;L_1,L_2;M]$ with $M = N_1-L_1+N_2-L_2$ is based on a Euclidean degree reduction of the second component numerator polynomials, while maintaining the degrees of the first component numerator polynomials at a constant value. We repeat that we continue to assume normality of the numerator polynomials in the sequences used. In this case, our sequence of SPAs has N_1, N_2 and L_1 kept fixed, and the abbreviated notation (3.2) is again helpful.

Initialisation Let $M_1 := N_1-L_1$. The entries

$$\begin{pmatrix} N_1-M_1 & M_1 \\ N_2 & \end{pmatrix}, \begin{pmatrix} N_1-M_1 & M_1+1 \\ N_2-1 & \end{pmatrix} \text{ and } \begin{pmatrix} N_1-M_1 & M_1+2 \\ N_2-2 & \end{pmatrix}$$

are needed to initialise the algorithm. The first of these comprises the numerator and denominator of a Padé approximant of series 1, and a Maclaurin polynomial of series 2. The other two entries would need special construction, by the Knight's Move algorithm for instance.

Iteration and Termination These are based on the recurrence

$$(3.9) \quad \begin{pmatrix} N_1-M_1 & M_1+3 \\ N_2-j-3 & \end{pmatrix} = (\alpha_j+x\beta_j)\begin{pmatrix} N_1-M_1 & M_1+2 \\ N_2-j-2 & \end{pmatrix}$$

$$+ (\gamma_j+x\delta_j)\begin{pmatrix} N_1-M_1 & M_1+1 \\ N_2-j-1 & \end{pmatrix} + \begin{pmatrix} N_1-M_1 & M_1 \\ N_2-j & \end{pmatrix}, \; j = 0, 1, \ldots, N_2-L_2-3.$$

The Euclidean interpretation of (3.9) allows sequential calculation of δ_j, β_j, γ_j and α_j for normal numerator polynomials. The value of δ_j is chosen to secure cancellation of the leading terms (of degree N_2-j)

in the second component; the value of β_j secures cancellation of the leading terms (of degree N_1-M_1+1) in the first component. The values of γ_j and α_j secure cancellation at orders x^{N_2-j-1} and x^{N_2-j-2} in the second component. The derivation of (3.9) follows repeated substitution of identity J, expressed with various indices, into identity Y.

4. Conclusions and Outlook

In the previous two sections, we presented three algorithms of Euclidean type for the construction of sequences of SPAs. Each of these algorithms fails if the relevant numerator polynomials do not have their full indicated degree. We mention that several similar algorithms have been found [21] for similar sequences. Our intention is to find reliable algorithms which work in degenerate cases as well. In the same way that Warner's modification converts Kronecker's algorithm into a reliable algorithm, so we have presented three algorithms which appear likely to have a similar modification. Unfortunately, replacement of the linear factors $(a_i x + b_i)$ in Fig.6 by polynomials of higher degree does not convert our first Kronecker type algorithm into a reliable one, and a more subtle modification is needed. We believe that such modifications exist, because, once the forms of the identities such as (3.5), (3.9) are known, they can be proved by induction. We also speculate that our algorithms can be speeded up by "divide and conquer" modifications. This might be useful in circumstances where the polynomials are known to have their full indicated order [4 and references therein; 12].

References

1. Aitken, A.C., Determinants and Matrices, Oliver and Boyd (1959).

2. Baker, G.A. Jr., Essentials of Padé Approximants, Academic Press (New York, 1975).

3. Baker, G.A. Jr. and Graves-Morris, P.R., Padé Approximants Part I in Encyclopedia of Mathematics, 13 (Addison-Wesley, 1981).

4. Brent, R.P., Gustavson, F.G. and Yun, D.Y.Y., Fast Solution of Toeplitz Systems of Equations and Computation of Padé Approximants, J. Algorithms 1, (1980), 259-295.

5. de Bruin, M.G., Generalised C-Fractions and a Multi-dimensional Padé Table, Thesis, Amsterdam University (1974).

6. de Bruin, M.G., Generalised Padé Tables and Some Algorithms Therein, in Proceedings of First French-Polish Meeting on Padé Approximation and Convergence Acceleration Techniques ed. J. Gilewicz CNRS (Marseille, 1982), 1-10.

7. de Bruin, M.G., Zeros of Polynomials Generated by 4-Term Recurrence Relations, in Rational Approximation and Interpolation eds. P.R. Graves-Morris, E.B. Saff and R.S. Varga, Springer (Heidelberg, 1984) 331-345.

8. Bultheel, A., Division Algorithms for Continued Fractions and the Padé Table, J. Comp. Appld. Math. 6, (1980), 259-266.

9. Bultheel, A. and van Barel, M., Padé Techniques for Model Reduction in Linear System Theory: a Survey, J. Comp. Appld. Math. 14, (1986), 401-438.

10. Gragg, W.B. and Lindquist, A., On the Partial Realisation Problem, Lin. Alg. and its Applcns. 50, (1983), 277-319.

11. Graves-Morris, P.R., The Numerical Calculation of Padé Approximants, in Padé Approximation and its Applications, ed. L. Wuytack, Springer (Heidelberg, 1980), 231-245.

12. Graves-Morris, P.R., Toeplitz Equations and Kronecker's Algorithm, C.S.S.P. 1, (1982), 289-304.

13. Graves-Morris, P.R., Vector-valued Rational Interpolants II, IMA J. Numer. Analy. 4, (1984), 209-224.

14. Kailath, T., Linear Systems, Prentice-Hall (1980).

15. Kalman, R.E., On Minimal Partial Realizations of a Linear Input/Output Map, in Aspects of Network and Systems Theory, eds. R.E. Kalman and N. de Claris, Holt, Reinhart and Winston (1971), 385-407.

16. McEliece, R.J. and Shearer, J.B., A Property of Euclid's Algorithm and an Application to Padé Approximation, SIAM J. Appl. Math. 34, (1978) 611-615.

17. MacWilliams, F.J. and Sloane, N.J.A., The Theory of Error Correcting Codes, North-Holland (1983).

18. Massey, J.L., Shift-Register Synthesis and BCH Decoding, IEEE Trans. Info. Theory IT-15, (1969), 122-127.

19. Padé, H., Sur la Généralisation des Fractions Continues Algébriques, J. de Math. 10 (1894), 291-329.

20. Warner, D., Hermite Interpolation with Rational Functions, Thesis, University of California, (1974).

21. Wilkins, J.M., Thesis, University of Kent at Canterbury, in preparation.

Postcript. Since original submission of this paper, we have become aware that J. van Iseghem has independently derived identity (B) in the same way. By combining it with the vector qd algorithm, she has derived an iterative construction of a different staircase sequence of denominator polynomials in the d-dimensional case [22].

22. van Iseghem, J., Thèse, Université de Lille, (1987); ICAM proceedings, (1987, in press).

ANALOGUES OF FREUD'S CONJECTURE FOR ERDöS TYPE WEIGHTS AND RELATED POLYNOMIAL APPROXIMATION PROBLEMS

by

Arnold Knopfmacher
Department of Mathematics, Witwatersrand University,
1 Jan Smuts Avenue, Johannesburg 2001, Republic of South Africa

and

D. S. Lubinsky
National Research Institute for Mathematical Sciences, CSIR,
P. O. Box 395, Pretoria 0001, Republic of South Africa

ABSTRACT. Let $W(x) := e^{-Q(x)}$, where $Q(x) \to \infty$ as $|x| \to \infty$ faster than any polynomial. Erdös [3] investigated orthogonal polynomials for weights of this type. Here we obtain asymptotics for the associated recurrence relation coefficients, analogous to those obtained recently for weights such as $\exp(-|x|^{\alpha})$, $\alpha > 0$. Our results apply to weights such as $W(x) := \exp(-\exp(|x|^{\alpha}))$ or $W(x) := \exp(-\exp(\exp(|x|^{\alpha})))$, $\alpha > 0$ arbitrary.

As a preliminary step, we investigate the possibility of approximation on the real line by weighted polynomials of the form $P_n(x)W(a_n x)$, where $P_n(x)$ is of degree at most n, and $\{a_n\}_1^{\infty}$ is a certain increasing sequence of positive numbers. Further, we investigate the asymptotic behaviour of entire functions that have nonnegative Maclaurin series coefficients, and that are associated with $W^2(x)$.

AMS (MOS) Classification: Primary 42C05, Secondary 41A10.
Keywords: Orthogonal Polynomials, Asymptotics, Recurrence Relation Coefficients, Freud's Conjecture, Erdös Weights.

1. Introduction

Let $W(x)$ be non-negative in $\mathbb{R}$, positive on a set of positive measure, and such that $W^2(x)$ has all power moments

$$\int_{-\infty}^{\infty} x^j W^2(x)dx , \quad j = 0,1,2,\ldots ,$$

finite. Associated with $W^2(x)$ are the orthonormal polynomials $\{p_n(W^2;x)\}_0^\infty$, satisfying

$$\int_{-\infty}^{\infty} p_n(W^2;x)p_m(W^2;x)W^2(x)dx = \begin{cases} 1, & m = n, \\ 0, & m \neq n. \end{cases}$$

The orthonormal polynomials satisfy a three term recurrence relation

$$xp_n(W^2;x) = A_{n+1}p_{n+1}(W^2;x) + B_n p_n(W^2;x) + A_n p_{n-1}(W^2;x) ,$$

where $B_n = B_n(W^2)$ is real and $A_n = A_n(W^2)$ is positive, $n = 1,2,3,\ldots$.

In this paper, we obtain asymptotics for A_n, B_n, as $n \to \infty$, for weights such as $W(x) := \exp(-Q(x))$, where $Q(x)$ is even and of faster than polynomial growth at infinity. More precisely, under various conditions on $Q(x)$, we show that

(1.1) $$\lim_{n \to \infty} A_n/Q^{[-1]}(n) = 1/2 ; \quad \lim_{n \to \infty} B_n/Q^{[-1]}(n) = 0 ,$$

where $Q^{[-1]}(x)$ denotes the inverse function of $Q(x)$, defined if $Q(x)$ is strictly increasing for large x. In particular, our results apply to weights such as $W^2(x) =: |x|^\rho \exp(-2Q(x))$, $\rho > -1$, where

$$Q(x) := \exp(\log(1+x^2)^{1+\alpha}) , \quad \alpha > 0 ,$$

or $$Q(x) := \exp(|x|^\alpha) , \quad \alpha > 0 ,$$

or $$Q(x) := \exp(\exp|x|^\alpha)) , \quad \alpha > 0 ,$$

and so on. Under somewhat weaker conditions on the weights than those considered here, Erdös [3] obtained asymptotics for the largest zeros of the orthogonal polynomials and showed that the contracted zero distribution is arcsine.

The methods used are similar to those of [10,15,16], which established asymptotics for A_n and B_n for a class of weights including

$$W^2(x) = |x|^\rho \exp(-|x|^\alpha) , \quad \rho > -1 , \quad \alpha > 0 ,$$

thereby establishing Freud's Conjecture in full generality. For α a positive even integer, the conjecture had earlier been proved by Magnus [18,19] and Magnus' results were sharpened to asymptotic expansions by Máté, Nevai, Zaslavsky [21] and Bauldry, Máté, Nevai [1].

As a preliminary step, we investigate the asymptotic behaviour of certain entire functions $G_Q(x)$ associated with $W^2(x)$, as well as of their derivatives. These functions were introduced in [12,13]. Next, we investigate approximation of functions on the real line by weighted polynomials of the form $P_n(x)W(Q^{[-1]}(n)x)$ where $P_n(x)$ is of degree n. In this connection, we remark that one can replace $Q^{[-1]}(n)$, by the positive root a_n of the equation

$$n = (2/\pi) \int_0^1 a_n t Q'(a_n t) dt / \sqrt{1-t^2} ,$$

whenever this is defined. This is so because

$$Q^{[-1]}(n)/a_n \to 1 , \quad n \to \infty ,$$

for the weights treated here. By contrast, this last limit relation is not true when $Q(x)$ is of polynomial growth at infinity, and in [15, 16] it was essential to consider $P_n(x)W(a_n x)$. The number a_n was introduced in [23].

We expect that the results and methods here will ultimately lead to strong asymptotics for the leading coefficients of orthonormal polynomials in the same way that the resolution of Freud's Conjecture led to strong asymptotics [17]. Further, we note that much as in [16], the results here yield upper bounds for Christoffel functions, and inequalities for the spacing of successive zeros of orthogonal polynomials. However, we omit the details, and refer the reader to [16] for results and proofs in a related context. See [8,32,33] for applications to asymptotics of orthogonal polynomials.

This paper is organised as follows: In section 2, we state our results on asymptotics of recurrence relation coefficients. In section 3, we state our results on asymptotics of certain entire functions, and related polynomial approximation problems. In section 4, we state and prove an infinite-finite range inequality. In section 5, we prove some of the results of section 3, namely, asymptotics of certain entire functions and their partial sums. In section 6, we establish the polynomial approximation results of section 3. Finally, in section 7, we prove the results of section 2.

2. Asymptotics for Recurrence Relation Coefficients

In this section, we state our results on asymptotics of recurrence relation coefficients for Erdös type weights. First we recapitulate on our notation. Given a function $w(x)$ non-negative in $\mathbb{R}$, positive on a set of positive measure, and with all power moments $\int_{-\infty}^{\infty} x^j w(x)dx$, $j = 0,1,2,\ldots$, finite, we call $w(x)$ a <u>weight</u> <u>function</u>. Associated with w is the sequence of orthonormal polynomials $\{p_n(w;x)\}_0^\infty$ where $p_n(w;x)$ has degree n,

$$(2.1)\qquad p_n(w;x) = \gamma_n(w)x^n + \ldots, \quad \gamma_n(w) > 0, \quad n = 0,1,2,\ldots,$$

and

$$(2.2)\qquad \int_{-\infty}^{\infty} p_n(w;x)p_m(w;x)w(x)dx = \begin{cases} 1, & m = n \\ 0, & m \neq n. \end{cases}$$

The orthonormal polynomials satisfy the recurrence relation

$$(2.3)\qquad xp_n(w;x) = A_{n+1}p_{n+1}(w;x) + B_n p_n(w;x) + A_n p_{n-1}(w;x),$$

where

$$(2.4)\qquad A_n := A_n(w) = \gamma_{n-1}(w)/\gamma_n(w),$$

and

$$(2.5)\qquad B_n := B_n(w) = \int_{-\infty}^{\infty} xp_n^2(w;x)w(x)dx,$$

$n = 1,2,3,\ldots$.

We shall find it convenient, as did G Freud, to formulate results for weights $w(x)$ of the form $W^2(x)$, where $W(x)$ is a non-negative function.

In the case when $W(x) := \exp(-Q(x))$, where $Q(x)$ is of polynomial growth at infinity, the asymptotic behaviour of $A_n(W^2)$ and $B_n(W^2)$ was investigated by Lubinsky, Mhaskar and Saff [14] - see also [5,10,18,19,20,27]. For example, for $W(x) := \exp(-|x|^\alpha)$, $\alpha > 0$, the results of [14] show that for $\rho > -1$,

$$\lim_{n\to\infty} A_n(|x|^\rho W^2)/n^{1/\alpha} = c_\alpha > 0 ,$$

and

$$\lim_{n\to\infty} B_n(|x|^\rho W^2)/n^{1/\alpha} = 0 .$$

thereby establishing a conjecture of Freud, special cases of which had earlier been established by Freud [5] ($\alpha = 4,6$) and Magnus [18,19] (α a positive even integer). The results of [15] do not apply to weights $W(x) := e^{-Q(x)}$, where $Q(x)$ is of faster than polynomial growth at ∞, for example $Q(x) = \exp(\log(1+x^2)^{1+\alpha})$ or $Q(x) = \exp(|x|^\alpha)$.

Erdös [3] established asymptotics for the largest zero of orthogonal polynomials associated with weights of this type, and also showed that the contracted zero distribution is arcsine. There are results that relate the asymptotic behaviour of the recurrence relation coefficients to that of the largest zeros - see the posthumously published paper [7] of Freud. However, it is not obvious that their hypotheses are satisfied by the weights considered here: so we adopt the approach of [15,16].

Throughout this paper, we write

$$f(x) \sim g(x)$$

if there exist positive constants C_1, C_2 such that

$$C_1 \le f(x)/g(x) \le C_2 ,$$

for the relevant range of x. Similar considerations apply to sequences and sequences of functions. Further, throughout, $C, C_1, C_2, \ldots$, denote positive constants independent of n and x. In different occurrences, the same symbol may denote different constants. We can now define a suitable class of weights:

Definition 2.1

Let $W(x) := e^{-Q(x)}$, where $Q(x)$ is even and continuous in $\mathbb{R}$, and $Q'''(x)$ exists for x large enough, while

(2.6) $\quad Q'(x) > 0\ ,\quad Q''(x) > 0\ ,\quad x \in [C_1,\infty)\ ,$

and

(2.7) $\quad Q''(x)/Q'(x) \sim Q'(x)/Q(x)\ ,\quad x \in [C_1,\infty)\ ,$

and

(2.8) $\quad |Q'''(x)/Q''(x)| \le C_2 Q'(x)/Q(x)\ ,\quad x \in [C_1,\infty)\ .$

Finally, assume that

(2.9) $\quad \lim_{x \to \infty} xQ'(x)/Q(x) = \infty\ ,$

and for each $\epsilon > 0$,

(2.10) $\quad xQ'(x)/Q(x) = O(Q'(x)^{\epsilon})\ ,\quad x \to \infty\ .$

Then we shall call $W(x)$ a very smooth Erdös weight and write $W \in VSE$.

Some remarks are in order concerning the above rather technical definition. The restrictions (2.7) and (2.8) are smoothness restrictions, while (2.9) ensures that $Q(x)$ grows faster than any polynomial at ∞ - by contrast, the weights of [15] typically satisfy

$$\lim_{x \to \infty} xQ'(x)/Q(x) = \alpha \in (0,\infty)\ ,$$

ensuring that $Q(x)$ grows approximately like $|x|^{\alpha}$ as $|x| \to \infty$. Condition (2.10) is a rather weak regularity condition. In fact, it is not difficult to show that for each $\epsilon > 0$,

$$\mathcal{E}_r := \{x \ge r : xQ'(x)/Q(x) \ge \epsilon(\log Q'(x))^{1+\epsilon}\}$$

satisfies

$$\int_{\mathcal{E}_r} \frac{dx}{x} \to 0\ ,\quad r \to \infty\ ,$$

under conditions (2.6) and (2.9).

As examples of $W \in VSE$, we mention $W := e^{-Q}$,

where $Q(x) := \exp(\{\log(1+x^2)\}^{1+\alpha})$, $\alpha > 0$,

or $Q(x) := \exp(|x|^{\alpha})$, $\alpha > 0$,
or $Q(x) := \exp(\exp(|x|^{\alpha}))$, $\alpha > 0$,

and so on.

In describing the asymptotics for the recurrence relation coefficients, we shall use the inverse function $Q^{[-1]}(\cdot)$, satisfying

(2.11) $Q^{[-1]}(Q(x)) = x$, x large enough .

The existence of $Q^{[-1]}$ follows from the fact that Q(x) is strictly increasing, x large enough. An alternative description can be given in terms of a_n, the root of

(2.12) $$n = \frac{2}{\pi}\int_0^1 a_n t Q'(a_n t)dt/\sqrt{1-t^2} ,$$

introduced by Mhaskar and Saff [23] in their investigation of sharp infinite-finite range inequalities. E.B. Saff has informed the authors that the number also appears in a dissertation of E.A. Rahmanov [30]. For the weights considered in this paper,

$$a_n/Q^{[-1]}(n) \to 1 , \quad n \to \infty ,$$

and so we prefer the simpler description involving $Q^{[-1]}$. However, we note that our proofs make extensive use of a_n and its associated theory.

As in [15], we can allow the weight to have infinities and zeros, and accordingly define a generalized Jacobi factor, similar to the generalized Jacobi weights on [-1,1] considered by Nevai [26].

<u>Definition 2.2</u> <u>Let</u>

(2.13) $$w_F(x) := \prod_{j=1}^{N} |x-z_j|^{\Delta_j} , \quad x \in \mathbb{R} ,$$

<u>where</u> $N \geq 1; z_1, z_2, \ldots z_N$ <u>are distinct complex numbers</u>, $\Delta_1, \Delta_2, \ldots, \Delta_N \in \mathbb{R}$ <u>and for each real</u> z_j <u>the corresponding</u> Δ_j <u>is larger than</u> $-1/2$. <u>Then we shall call</u> $w_F(x)$ <u>a generalized Jacobi factor</u>.

One of our main results is:

<u>Theorem 2.3</u>

<u>Let</u> $W(x) := e^{-Q(x)} \in \text{VSE}$, $w_F(x)$ <u>be a generalized Jacobi factor</u>, $\varphi(x) \in L_\infty(\mathbb{R})$ <u>and</u> $\varphi(x)$ <u>be non-negative with</u>

$$(2.14) \qquad \lim_{|x|\to\infty} \varphi(x) = 1 .$$

<u>Further, let</u> $g(x)$ <u>be a function continuously differentiable in</u> $\mathbb{R}$, <u>with</u>

$$(2.15) \qquad \max\{|g'(x)| : |x| \leq r\} = O(Q(r)^{1/2}/(r\log r)) , \quad r \to \infty .$$

<u>Finally, let</u>

$$(2.16) \qquad \hat{W}(x) := W(x) w_F(x) \varphi(x) e^{g(x)} , \quad x \in \mathbb{R} .$$

<u>Then</u>

$$(2.17) \qquad \lim_{n\to\infty} A_n(\hat{W}^2)/Q^{[-1]}(n) = 1/2 ,$$

<u>and</u>

$$(2.18) \qquad \lim_{n\to\infty} B_n(\hat{W}^2)/Q^{[-1]}(n) = 0 .$$

Note that for large $|x|$, one may rewrite (2.16) in the form

$$(2.19) \qquad \hat{W}(x) = \prod_{j=1}^{N} |x-z_j|^{\Delta_j} \exp(-Q(x)+g(x)+o(1)) , \quad |x| \to \infty .$$

When $Q(x)$ grows faster than $\exp(|x|^\alpha)$, some $\alpha > 1$, we can prove (2.17) and (2.18) without assuming Q'' or Q''' exist, but shan't pursue this here. The freedom of choice of $\varphi(x)$ does allow $\hat{W}(x)$ to have several disjoint intervals of support. We suspect that one should be able to replace 1/2 in (2.15) by 1. At least for entire $g(x)$, we can allow faster growth than that in (2.15):

<u>Theorem 2.4</u>

Let $W(x), w_F(x)$ and $\varphi(x)$ be as in Theorem 2.3, and let $g(x)$ be a real entire function such that for some $0 < \delta < 1$,

$$(2.20) \quad \max\{|g(t)| : |t| \le r, \; t \in \mathbb{C}\} = O(Q(r)^{1-\delta}), \quad r \to \infty .$$

Then if $\hat{W}(x)$ is given by (2.16), the conclusions (2.17) and (2.18) of Theorem 2.3 remain valid.

Both Theorem 2.3 and 2.4 follow by setting $V := e^g$ in the following result:

Theorem 2.5 Let $W(x)$, $w_F(x)$ and $\varphi(x)$ be as in Theorem 2.3, and let $V(x)$ be a non-negative measurable function on $\mathbb{R}$ with the following properties: $V(x)$ is bounded above in each finite interval;

$$(2.21) \quad \lim_{|x|\to\infty} \log V(x)/Q(x) = 0 ;$$

For some fixed sequence of non-negative numbers $\{c_n\}_1^\infty$ satisfying

$$(2.22) \quad \lim_{n\to\infty} c_n/Q^{[-1]}(n) = 1 ,$$

and

$$(2.23) \quad \lim_{n\to\infty} Q(c_n)/n = \infty ,$$

there exist for n large enough, a polynomial $U_n(x)$ of degree at most $n/2$, such that

$$(2.24) \quad \lim_{n\to\infty} |V(c_n x)U_n(x)| = 1 \quad \text{a.e. in } (-1,1) ,$$

and

$$(2.25) \quad \|V(c_n x)U_n(x)\|_{L_\infty[-1,1]} \le C , \quad n \text{ large enough} .$$

Then if

$$(2.26) \quad \hat{W}(x) := W(x)w_F(x)\varphi(x)V(x) , \quad x \in \mathbb{R} ,$$

the conclusions (2.17) and (2.18) of Theorem 2.3 remain valid.

3. Entire Functions, Weights and Weighted Polynomial Approximations

In this section, we state some results on entire functions introduced in [12,13]. Define

$$(3.1) \qquad G_Q(x) := 1 + \sum_{n=1}^{\infty} (x/q_n)^{2n} e^{2Q(q_n)} n^{-1/2} ,$$

where for n large enough, q_n is Freud's quantity [6], the positive root of the equation

$$(3.2) \qquad n = q_n Q'(q_n) .$$

When $W := e^{-Q}$ satisfies (2.6), it is easy to see that q_n is uniquely defined for n large enough, and increases to ∞ as n increases to ∞. For those n for which q_n is not defined, we can simply set $q_n := 1$. The entirety of $G_Q(x)$ is established in section 5. When Q(x) is of smooth polynomial growth at ∞, the asymptotic behaviour of $G_Q(x)$ was obtained in [12], and for Q(x) of faster than polynomial growth at ∞, the asymptotics appear in [13]. Here we obtain sharper asymptotics than in [13] and under simpler conditions on W(x). We also investigate the asymptotic behaviour of the derivatives:

<u>Theorem 3.1</u> <u>Let</u> $W(x) := e^{-Q(x)}$, <u>where</u> Q(x) <u>is even and continuous in</u> $\mathbb{R}$, <u>and</u> $Q'''(x)$ <u>exists for</u> x <u>large enough, while (2.6), (2.7) and (2.8) are satisfied. Then for each fixed non-negative integer</u> j, <u>we have</u>

$$(3.3) \quad G_Q^{(j)}(x) = (2Q'(x))^j e^{2Q(x)} (\pi T(x))^{1/2} \{1+O(\{\log Q(x)\}^{3/2} Q(x)^{-1/2})\}, \quad |x| \to \infty ,$$

<u>where</u>

$$(3.4) \qquad T(x) := 1 + xQ''(x)/Q'(x) , \quad |x| \ \underline{\text{large enough}} .$$

In the proof of Freud's Conjecture [15] an important role was played by certain weighted polynomial approximations [16]. The results of [16] also resolved a conjecture of Saff [31] in certain cases. For $W \in VSE$ we obtain analogous results here, though the proofs are somewhat simpler than in [16] - the partial sums of $G_Q(x)$ provide good enough approximations, and so the Lagrange interpolation and delicate contour integral estimations of [16] are no longer required.

Following we estimate the approximation power of the partial sums of $G_Q(x)$:

Theorem 3.2 Let $W(x) := e^{-Q(x)}$, where $Q(x)$ is even and continuous in $\mathbb{R}$, and $Q'''(x)$ exists for x large enough, while (2.6), (2.7), (2.8) and (2.10) are satisfied. Let j be a fixed non-negative integer, and let $S_{2n}(x)$ denote the (n+1)th partial sum of $G_Q(x)$, so that $S_{2n}(x)$ has degree 2n; n = 0,1,2,... . Let $\varepsilon > 0$. Then for n large enough,

$$(3.5) \qquad \lim_{n \to \infty} \| 1 - S_{2n}^{(j)}(x)/G_Q^{(j)}(x) \|_{L_\infty(|x| \le q_{n(1-n^{-1/2+\varepsilon})})} = 0 .$$

Using the above theorem, we prove an analogue of Theorem 2.4 in Lubinsky and Saff [16] for $W \in VSE$. We remark as for Theorem 2.3, that when Q(x) grows faster than $\exp(|x|^\alpha)$, some $\alpha > 1$, the restrictions on Q' and Q''' can be dropped.

Theorem 3.3 Let $W(x) := e^{-Q(x)} \in VSE$. Let g(x) be continuous in $\mathbb{R}$, $0 < \delta \le 1$, and $\{c_n\}_1^\infty$ be a sequence of positive numbers satisfying

$$(3.6) \qquad \lim_{n \to \infty} c_n / Q^{[-1]}(n) = 1 .$$

Then the following are equivalent:

(i) There exist polynomials $P_n(x)$ of degree at most δn, n large enough, such that

$$(3.7) \qquad \lim_{n \to \infty} \| g(x) - P_n(x) W(c_n x) \|_{L_\infty(\mathbb{R})} = 0 .$$

(ii) $g(x) = 0 , \quad |x| \ge 1 .$

We note that in particular, one may choose $c_n = q_n$ or q_{Ln}, L any fixed positive number, n large enough. Further, whenever a_n defined by

(2.12) exists for n large enough (true if, for example, we make mild additional assumptions on $Q'(x)$ near 0), then one may choose $c_n = a_n$, n large enough.

4. An Infinite-Finite Range Inequality

In this section, we prove an infinite-finite range inequality similar to those in Lubinsky, Mhaskar and Saff [15]. See Mhaskar and Saff [24,25], Nevai [28] and [2,11,16] for discussions from different viewpoints of infinite-finite range inequalities. We use $\mathcal{P}_n$ to denote the class of polynomials of degree at most n:

Theorem 4.1

Let $W(x) := e^{-Q(x)} \in VSE$, except that we do not assume (2.10). Let $0 < p \leq \infty$ and $\hat{W}(x)$ be a non-negative function such that

$$(4.1)\qquad x^n\hat{W}(x) \in L_p(\mathbb{R}),\quad n = 0,1,2,\ldots,$$

and

$$(4.2)\qquad \limsup_{|x|\to\infty}\{\log 1/\hat{W}(x)\}/Q(x) < \infty.$$

Then there exist respectively increasing and decreasing sequences $\{c_n\}_1^\infty$ and $\{\delta_n\}_1^\infty$ of positive numbers, with

$$(4.3)\qquad \lim_{n\to\infty} c_n/q_n = 1\ ;$$

$$(4.4)\qquad \lim_{n\to\infty} \delta_n = 0\ ;$$

and for $n = 1,2,3,\ldots$, and each $P \in \mathcal{P}_n$,

$$(4.5)\qquad \|P\hat{W}\|_{L_p(|x|\geq c_n)} \leq \delta_n \|P\hat{W}\|_{L_p(|x|\leq c_n)},$$

and

$$(4.6)\qquad \|P\hat{W}\|_{L_p(\mathbb{R})} \leq (1+\delta_n)\|P\hat{W}\|_{L_p(|x|\leq c_n)}.$$

Furthermore, we may choose $\{c_n\}_1^\infty$ to be any sequence satisfying (4.3) and

$$(4.7)\qquad \lim_{n\to\infty} Q(c_n)/n = \infty .$$

We remark that in the case that $\hat{W} = W$ and Q is convex in all of $(0,\infty)$, a more precise inequality appears in [15, Theorem 2.6] in terms of a_n, the root of (2.12), which should probably be called the Mhaskar-Rahmanov-Saff or Mhaskar-Saff number. We now establish some basic technical properties of W:

Lemma 4.2

Let $W(x) := e^{-Q(x)} \in$ VSE, except that we do not assume (2.10). Then the following hold:

(i) For each $K > 0$, we have for x large enough,

$$(4.8)\qquad Q'(x) \geq x^K .$$

(ii) For each $L > 1$,

$$(4.9)\qquad \lim_{x\to\infty} Q'(Lx)/Q'(x) = \infty .$$

(iii) For each $L > 1$,

$$(4.10)\qquad \lim_{x\to\infty} Q(Lx)/Q(x) = \infty .$$

(iv) Uniformly in closed subsets of $(1,\infty)$,

$$(4.11)\qquad \lim_{n\to\infty} Q(q_n x)/n = \infty ,$$

and uniformly in compact subsets of $[0,1]$,

$$(4.12)\qquad \lim_{n\to\infty} Q(q_n x)/n = 0 .$$

In particular,

(4.13) $$\lim_{n \to \infty} Q(q_n)/n = 0 .$$

(v) <u>For each</u> $L > 0$,

(4.14) $$\lim_{n \to \infty} q_{Ln}/q_n = 1 .$$

(vi) <u>Given</u> $\epsilon > 0$, <u>we have for</u> m, n <u>large enough with</u> $m \geq n$,

(4.15) $$q_m/q_n = O(m/n)^{\epsilon} .$$

(vii) <u>For each</u> $L > 0$,

(4.16) $$\lim_{n \to \infty} q_n/Q^{[-1]}(Ln) = 1 .$$

(viii) <u>If also</u> $Q'(x)$ <u>exists in all of</u> $(0,\infty)$ <u>and</u> $xQ'(x)$ <u>remains bounded as</u> $x \to 0+$, <u>and</u> a_n <u>denotes for n large enough the positive root of the equation (2.12), then</u> a_n <u>exists and is unique for</u> n <u>large enough and</u>

(4.17) $$\lim_{n \to \infty} a_n/q_n = 1 .$$

<u>Proof</u>

(i) Now if $K > 0$ is given and x is large enough, say $x \geq C$,

$$Q'(x)/Q'(C) = \exp\left(\int_C^x Q''(u)/Q'(u)du\right)$$

$$\geq \exp\left(K \int_C^x du/u\right) \text{ (by (2.7) and (2.9))}$$

$$= (x/C)^K .$$

Then (4.8) follows.

(ii) Much as in (i) if x is large enough, and $K > 0$,

$$Q'(Lx)/Q'(x) = \exp\left(\int_x^{Lx} Q''(u)/Q'(u)du\right)$$

$$\geq \exp\left(K\int_x^{Lx} du/u\right) \quad \text{(by (2.7) and (2.9))}$$

$$= L^K .$$

Since K may be made arbitrarily large, (4.9) follows.

(iii) This is similar to (ii).

(iv) Let $\epsilon > 0$. Since $Q(u)$ is strictly increasing for large enough u, we have for $x \geq 1 + 2\epsilon$ and n large enough,

$$\begin{aligned} Q(q_n x) &\geq Q(q_n(1+2\epsilon)) \\ &= Q(q_n(1+\epsilon)) + q_n \epsilon Q'(q_n \xi) , \end{aligned}$$

where ξ lies between $1 + \epsilon$ and $1 + 2\epsilon$. Since $Q'(u)$ is strictly increasing for large enough u, we have

$$\begin{aligned} Q(q_n x)/n &\geq q_n \epsilon Q'(q_n(1+\epsilon))/n \\ &= \epsilon Q'(q_n(1+\epsilon))/Q'(q_n) \quad \text{(by (3.2))} \\ &\to \infty , \quad \text{as } n \to \infty , \end{aligned}$$

by (4.9).

Next, if $x \in [0,1]$, the monotonicity of $Q(u)$ for large u, and its continuity in $\mathbb{R}$, yield for large n,

$$\begin{aligned} Q(q_n x)/n &\leq Q(q_n)/n \\ &= Q(q_n)/(q_n Q'(q_n)) \\ &\to 0 , \quad n \to \infty , \end{aligned}$$

by (2.9).

(v) By the definition (3.2) of q_n,

$$L = q_{LN}Q'(q_{LN})/(q_nQ'(q_n)) .$$

If $\liminf_{n \to \infty} q_{Ln}/q_n < 1$ or $\limsup_{n \to \infty} q_{Ln}/q_n > 1$, then (4.9) easily yields a contradiction.

(vi) If $m \geq n$ and m,n are large enough,

$$m/n = (q_mQ'(q_m))/(q_nQ'(q_n)) = \exp(\int_{q_n}^{q_m} \frac{d}{du} \log\{uQ'(u)\}du)$$

$$= \exp(\int_{q_n}^{q_m} [\frac{1}{u} + \frac{Q''(u)}{Q'(u)}]du) .$$

Proceeding as in the proof of (i), we may use (2.7) and (2.9) to deduce that for each $K > 0$ and $m \geq n$ large enough,

$$m/n \geq \exp(K\int_{q_n}^{q_m} du/u)$$
$$= (q_m/q_n)^K .$$

Setting $K = 1/\epsilon$, we obtain (4.15).

(vii) By (2.9), for n large enough,

$$n = q_nQ'(q_n) > Q(q_n)$$

and so

$$Q^{[-1]}(n) > q_n .$$

If on the other hand, we have for some $\epsilon > 0$ and infinite sequence $\mathcal{S}$ of positive integers,

$$Q^{[-1]}(n) \geq (1+2\epsilon)q_n , \quad n \in \mathcal{S} ,$$

then

$$n \geq Q((1+2\epsilon)q_n) , \quad n \in \mathcal{S} ,$$

contradicting (4.11). Thus

$$\limsup_{n \to \infty} Q^{[-1]}(n)/q_n \le 1 \ ,$$

and (4.16) follows for L = 1. The case of general L follows from (4.14).

(viii) The existence and uniqueness of a_n for n large, follows as in Lemma 3.2 of [16]. Next, for n large enough,

$$\begin{aligned} q_n Q'(q_n) = n &= (2/\pi) \int_0^1 a_n t Q'(a_n t)/\sqrt{1-t^2}\,dt \\ &\le a_n Q'(a_n) \ , \end{aligned}$$

by monotonicity of $xQ'(x)$, x large. Hence

(4.17) $\quad q_n \le a_n \ , \quad$ n large enough .

Next if $0 < \epsilon < 1$ is fixed and n is large enough, we have

$$n \ge (2/\pi) \int_{1-\epsilon}^1 a_n t Q'(a_n t)/\sqrt{1-t^2}\,dt + O(1/a_n) \ ,$$

so that

$$n/2 \ge a_n(1-\epsilon)Q'(a_n(1-\epsilon)) \ (2/\pi)\int_{1-\epsilon}^1 dt/\sqrt{1-t^2} \ .$$

Hence for some $L = L(\epsilon)$,

$$Ln \ge a_n(1-\epsilon)Q'(a_n(1-\epsilon))$$

and so

$$q_{Ln} \ge a_n(1-\epsilon) \ .$$

Then (4.14), the arbitrariness of ϵ, and (4.17) yield the result. □

<u>Proof of Theorem 4.1</u>

First note the well known inequality [4, p. 118],

$$|P(x)| \le T_n(|x|)\|P\|_{L_\infty[-1,1]}, \quad |x| \ge 1, \quad P \in \mathcal{P}_n,$$

where $T_n(x)$ is the classical Chebyshev polynomial of degree n. By a linear transformation, we deduce that for $P \in \mathcal{P}_n$,

$$\begin{aligned} |P(x)| &\le T_n(|4x-3|)\|P\|_{L_\infty[1/2,1]} \\ &\le (14|x|)^n\|P\|_{L_\infty[1/2,1]}, \quad |x| > 1, \\ &\le (15|x|)^n\|P\|_{L_p[1/2,1]}, \quad |x| > 1, \end{aligned}$$

provided n is large enough. Here we have used standard Nikolskii inequalities for a finite interval [26, p. 114]. A scale change then yields, for n large enough,

$$(4.18) \qquad |P(x)| \le (15|x|/q_n)^n\|P\|_{L_p[q_n/2,q_n]}, \quad |x| > q_n,$$

$P \in \mathcal{P}_n$. Next, in view of (4.2), there exists $K \ge 2$ and $C > 0$ such that

$$(4.19) \qquad W^K(x) \le \hat{W}(x) \le W^{1/K}(x), \quad |x| \ge C.$$

Further for $x \in [q_n/2, q_n]$,

$$\begin{aligned} \hat{W}(x)^{-1/n} &\le W^{-(K/n)}(x) \le \exp(KQ(q_n)/n) \\ &\to 1, \quad n \to \infty, \end{aligned}$$

by (4.13). Hence from (4.18), we have for n large enough,

$$|P(x)| \le (16|x|/q_n)^n\|P\hat{W}\|_{L_p[q_n/2,q_n]},$$

$|x| > q_n$, $P \in \mathcal{P}_n$. Then by (4.19), we have for $r \ge 1$,

$$(4.20) \qquad \|P\hat{W}\|_{L_p(|x|\ge rq_n)} \le \|PW^{1/K}\|_{L_p(|x|\ge rq_n)}$$

$$\le (16/q_n)^n \|P\hat{W}\|_{L_p[q_n/2,q_n]} \|x^n W^{1/K}(x)\|_{L_p(|x|\ge rq_n)} .$$

Now by differentiating $x^{2n}W^{1/K}(x)$, we see that its maximum is attained when $x = q_{2Kn}$. Thus for some C,

$$|x^n W^{1/K}(x)| \le q_{2Kn}^{2n} \; W^{1/K}(q_{2Kn})|x|^{-n} , \quad |x| \ge C ,$$

and so for n large enough,

$$\{(16/q_n)^n \|x^n W^{1/K}(x)\|_{L_p(|x|\ge 20q_n)}\}^{1/n}$$

$$\le 16(q_{2Kn}^2/q_n)W^{1/(Kn)}(q_{2Kn})\|x^{-n}\|^{1/n}_{L_p(|x|\ge 20q_n)}$$

$$\le 16q_{2Kn}(1+o(1))(1+o(1))(20q_n)^{-1}(1+o(1)) ,$$

by (4.14), (4.13) and (4.15) (with m large and n fixed there). Applying (4.14) again, we see that for some $0 < \rho < 1$ and n large enough,

$$(4.21) \qquad (16/q_n)^n \|x^n W^{1/K}(x)\|_{L_p(|x|\ge 20q_n)} \le \rho^n .$$

Next, if $r \ge 1$,

$$(4.22) \qquad \{(16/q_n)^n \|x^n W^{1/K}\|_{L_p(rq_n\le|x|\le 20q_n)}\}^{1/n}$$

$$\le (16/q_n)(20q_n)\exp(-Q(rq_n)/(Kn))(20q_n)^{1/(pn)} .$$

Now by (4.11), for each fixed $r > 1$,

$$\lim_{n\to\infty} Q(rq_n)/n = \infty ,$$

and so we can choose

$$r := 1 + \epsilon_n , \quad n \text{ large enough} ,$$

where $\{\epsilon_n\}_1^\infty$ is a decreasing sequence of positive numbers with limit 0, such that

$$c_n := rq_n := (1+\epsilon_n)q_n$$

satisfies

$$\lim_{n\to\infty} Q(c_n)/n = \infty ,$$

and

$$\lim_{n\to\infty} c_n/q_n = 1 .$$

Then (4.3), (4.7) hold and we see that the right hand side of (4.22) is bounded above by $\hat{\rho}^n$, for n large enough where $0 < \hat{\rho} < 1$. Combining (4.20) to (4.22), we obtain for some $0 < \rho' < 1$, n large enough, and all $P \in P_n$,

$$\|P\hat{W}\|_{L_p(|x|\ge c_n)} \le \rho'^n \|P\hat{W}\|_{L_p[q_n/2,q_n]} .$$

Then (4.5) and (4.6) follow. □

5. Proof of Theorems 3.1 and 3.2

In proving Theorem 3.1, we shall use Laplace's method, much as in [12] and this will require several lemmas:

Lemma 5.1

Let $W := e^{-Q}$ satisfy the hypotheses of Theorem 3.1, and let $T(x)$ be defined by (3.4). Then for some $C_1, C_2, C_3 > 0$ and $u \in (C_1,\infty)$,

(5.1) $Q''(u) \sim Q'(u)^2/Q(u)$,

(5.2) $uQ'(u)/Q(u) \ge C_2$,

(5.3) $T(u) \sim uQ'(u)/Q(u)$,

(5.4) $T(q_u) \sim u/Q(q_u)$,

(5.5) $T'(u) = O(u\{Q'(u)/Q(u)\}^2)$.

(5.6) $Q(q_u) = O(u)$,

(5.7) $q_u'/q_u = 1/(uT(q_u))$,

(5.8) $q_{2u} \geq q_u(1+C_3Q(q_u)/u)$.

<u>Proof</u>

Firstly (5.1) follows from (2.7). Next for u large enough, Q'(u) is increasing, so with a suitable C,

$$Q(u) = Q(C) + \int_C^u Q'(t)dt$$

$$\leq Q(C) + uQ'(u) \leq 2uQ'(u) .$$

This yields (5.2). Next, (5.3) is an immediate consequence of (2.6), (2.7), (3.4) and (5.2). Further, (5.4) follows from (5.3) and the definition (3.2) of q_n.

To prove (5.5), we compute T'(u) from (3.4):

$$\begin{aligned} T'(u) &= Q''(u)/Q'(u) + uQ\ (u)/Q'(u) - u(Q''(u)/Q'(u))^2 \\ &= O(Q'(u)/Q(u)) + O(u(Q'(u)/Q(u))^2) + O(u(Q'(u)/Q(u))^2), \end{aligned}$$

by (2.7) and (2.8). Then (5.5) follows.

To obtain (5.6), we replace u by q_u in (5.2). Next, differentiating the relation

$$q_uQ'(q_u) = u ,$$

we obtain

$$q_u'Q'(q_u)T(q_u) = 1 .$$

Upon applying (3.2), we then obtain (5.7).

Finally we prove (5.8). For large enough u, there exists v between u, 2u such that

$$\begin{aligned} q_{2u} &= q_u + uq_v' \\ &= q_u + uq_v/(vT(q_v)) \\ &\geq q_u + uCq_vQ(q_v)/v^2 \quad \text{(by (5.4))} \\ &\geq q_u + uCq_uQ(q_u)/(2u)^2 , \end{aligned}$$

by monotonicity of q_u and $Q(q_u)$. Then (5.8) follows. □

<u>Lemma 5.2</u>

<u>Let</u> j <u>be a fixed non-negative integer, let</u> $W = e^{-Q}$ <u>satisfy the hypotheses of Theorem 3.1, and let</u> $G_Q(u)$ <u>be defined by (3.1). Let</u>

$$(5.9) \qquad H(x,u) := j \log(2u/x) + \sum_{k=1}^{j-1} \log(1-k/(2u))$$

$$+ 2u \log(x/q_u) - (\tfrac{1}{2})\log u + 2Q(q_u) , \quad u > 0 ,$$

<u>where if</u> j = 0,1, <u>the sum with upper index</u> j - 1 <u>is interpreted as</u> 0. <u>Then, for some suitable constant</u> C

$$(5.10) \qquad G_Q^{(j)}(x) = C + \sum_{n>j/2} \exp(H(x,n)) .$$

<u>and if ' denotes partial differentiation with respect to</u> u <u>for</u> x <u>fixed</u>,

$$(5.11) \qquad H'(x,u) = (j-1/2)/u + \sum_{k=1}^{j-1} (ku^{-2}/2)(1-k/(2u))^{-1} + 2 \log(x/q_u).$$

$$(5.12) \qquad H''(x,u) = - (j-1/2)u^{-2} - \sum_{k=1}^{j-1} (ku^{-2}/2)^2(1-k/(2u))^{-2}$$

$$- \sum_{k=1}^{j-1} (ku^{-3})(1-k/(2u))^{-1} - 2q_u'/q_u \ .$$

and

(5.13) $$H''(x,u) = -2(uT(q_u))^{-1}(1+O(Q^{-1}(q_u))) \ .$$

Proof

Firstly, (5.10) follows from (3.1) and (5.9) by elementary manipulations. Next, (5.11) follows from (5.9), and (3.2), noting that

$$(Q(q_u))' = Q'(q_u)q_u' = uq_u'/q_u \ .$$

Next, (5.12) follows from (5.11), and we see then, using (5.4) and (5.7) that

$$H''(x,u) = -2(uT(q_u))^{-1} + O(u^{-2})$$
$$= -2(uT(q_u))^{-1}(1+O(Q(q_u)^{-1})) \ . \qquad \square$$

Lemma 5.3

Assume the notation and assumptions of Lemma 5.2. For x large enough, let $y = y(x)$ denote the root of the equation

(5.14) $$H'(x,y) = 0 \ .$$

(i) Then there exists $B > 0$ such that for x large enough, there is a unique $y \in (B,\infty)$ satisfying (5.14). Further, for large enough x,

(5.15) $$x = q_y \exp\{-(j-1/2)/(2y) - \sum_{k=1}^{j-1} (ky^{-2}/4)(1-k/(2y))^{-1}\}$$

(5.16) $$= q_y - q_y(j-1/2)/(2y) + O(q_y y^{-2}) \ ,$$

and

(5.17) $$q_{y/2} \le x \le q_{2y} \ .$$

(ii) For large enough x,

(5.18) $$Q(x) = Q(q_y) - (j-1/2)/2 + O(Q(x)^{-1}) \ ,$$

(5.19) $H(x,y) = j \log(2y/x) - (1/2)\log y + 2Q(x) + O(1/Q(x))$,

<u>and</u>

(5.20) $Q'(x) = Q'(q_y) + O(Q'(x)/Q(x))$.

(iii) <u>For large enough</u> x,

(5.21) $T(q_y) = T(x)(1+O(1/Q(x)))$,

<u>and</u>

(5.22) $y = xQ'(x)(1+O(1/Q(x)))$.

<u>Proof</u>

(i) Note first from (5.12) that $H''(x,u)$ is independent of x, and a function of u only. From (5.13), it follows that there exists B independent of x such that for all x,

$$H''(x,u) < 0 \ , \quad u \geq B \ .$$

Hence $H'(x,u)$ is strictly decreasing for $u \in (B,\infty)$ and all x large enough. Since from (5.11), for large enough x,

$$H'(x,B) > 0 \quad \text{and} \quad H'(x,\infty) = -\infty \ ,$$

there is a unique $y \in (B,\infty)$ satisfying (5.14).

Next, (5.15) and (5.16) follow immediately from (5.11).

To prove (5.17), we have from (5.16), and for x large enough,

$$q_y(1-C/y) \leq x \leq q_y(1+C/y) \ .$$

Using (5.8), we deduce for x large enough,

$$x \leq q_y(1+C_3Q(q_y)/y) \leq q_{2y} \ ,$$

and

$$x \geq q_{y/2}(1+C_3Q(q_{y/2})/(y/2))(1-C/y)$$

$\geq q_{y/2}$.

Then (5.17) follows.

(ii) To prove (5.18), we note that there exists v between x and q_y such that

$$Q(x) = Q(q_y) + (x-q_y)Q'(q_y) + (x-q_y)^2 Q''(v)/2$$

$$= Q(q_y)+\{-q_y(j-1/2)/(2y)+O(q_y y^{-2})\}Q'(q_y)+O(\{q_y/y\}^2 Q'(v)^2/Q(v))$$

(by (5.1), (5.16))

$$= Q(q_y)-(j-1/2)/2+O(y^{-1})+O(q_{2y}^2 y^{-2} Q'(q_{2y})^2/\max\{Q(x),Q(q_y)\})$$

$$= Q(q_y) - (j-1/2)/2 + O(y^{-1}) + O(1/\max\{Q(x),Q(q_y)\}) \ ,$$

where we have used (3.2) and (5.17). It follows that

$$(5.23) \qquad Q(x) \sim Q(q_y) \ ,$$

and in view of (5.6), we obtain (5.18).

Next, by (5.9),

$$H(x,y) = j \log(2y/x)+ \sum_{k=1}^{j-1} \log(1-k(2y))+2y \log(x/q_y)-(1/2)\log y+2Q(q_y).$$

Using (5.15) and (5.18), we deduce that

$$H(x,y) = j \log(2y/x)+O(y^{-1}) - (j-1/2)+O(y^{-1}) - (1/2)\log y + 2Q(x) + (j-1/2) + O(Q(x)^{-1})$$

Since

$$Q(x) \sim Q(q_y) = O(y) \ ,$$

this last relation yields (5.19).

Next, there exists v between x and q_y such that

$$\begin{aligned} Q'(x) &= Q'(q_y) + (x-q_y)Q''(v) \\ &= Q'(q_y) + O(q_y/y)Q'(v)^2/Q(v)) \\ &\quad \text{(by (5.16) and (5.1))} \\ &= Q'(q_y) + O(q_y y^{-1} Q'(q_{2y})^2/Q(x)) \\ &\quad \text{(by (5.17) and (5.23))} \\ &= Q'(q_y) + O(Q'(q_{2y})/Q(x)) \ . \end{aligned}$$

Here

$$\begin{aligned} Q'(q_{2y}) = 2y/q_{2y} &= O(y/q_y) \\ &= O(Q'(q_y)) \ . \end{aligned}$$

Thus

$$Q'(x) = Q'(q_y)\{1+O(1/Q(x))\} \ .$$

Then (5.20) follows.

(iii) There exists v between x and q_y such that

$$\begin{aligned} T(x) &= T(q_y) + (x-q_y)T'(v) \\ &= T(q_y) + O(q_y y^{-1} q_y Q'(q_y)^2/Q(q_y)^2) \\ &\quad \text{(by (5.16), (5.5), (5.18) and (5.20))} \\ &= T(q_y) + O(y/Q(q_y)^2) \\ &= T(q_y)(1+O(1/Q(q_y))) \ , \end{aligned}$$

by (5.4). Then (5.21) follows.

Finally, by (5.16) and (5.20),

$$\begin{aligned} y &= q_y Q'(q_y) \\ &= (x+O(q_y/y))Q'(q_y) \\ &= x(Q'(x) + O(Q'(x)/Q(x))) + O(1) \\ &= xQ'(x) + O(xQ'(x)/Q(x)) \ . \end{aligned}$$

in view of (5.2). □

Lemma 5.4

Assume the notation and assumptions of Lemma 5.2. Let $K > 0$ and let y be as in Lemma 5.3. Let

$$(5.24)\qquad w := w(y) := K(\log Q(q_y))^{1/2} y Q(q_y)^{-1/2}.$$

Then if K is large enough, we have as $x \to \infty$,

$$(5.25)\qquad I(x) := \int_{y-w}^{y+w} \exp(H(x,u))du$$

$$(5.26)\qquad = (2Q'(x))^j e^{2Q(x)} (\pi T(x))^{1/2}\{1+O(\{\log Q(x)\}^{3/2} Q(x)^{-1/2})\}.$$

Proof

Let $u \in [y-w, y+w]$. Then there exists v between u and y such that

$$H(x,u) = H(x,y) + (u-y)H'(x,y) + (u-y)^2 H''(x,v)/2 .$$

Since (5.14) holds and $w = o(y)$, we deduce with the aid of (5.4) and (5.13) that

$$(5.27)\qquad H(x,u) = H(x,y) - (u-y)^2 (vT(q_v))^{-1} + O(w^2 y^{-2}).$$

Here

$$(vT(q_v))^{-1} - (yT(q_y))^{-1} = \int_y^v \frac{d}{du}\{uT(q_u)\}^{-1} du$$

$$= \int_y^v \{-u^{-2}T(q_u)^{-1} - u^{-1}T'(q_u)q_u' T^{-2}(q_u)\}du$$

$$= O(\max\{Q(q_u) : u \in [y-w, y+w]\} y^{-3} w),$$

by (5.4), (5.5) and (5.7). Now if $s, t \in [y-w, y+w]$, there exists z between s and t such that

$$Q(q_s) - Q(q_t) = (s-t)Q'(q_z)q_z'$$
$$= O(wQ(q_z)/z) \qquad \text{(by (3.2), (5.4) and (5.7))}$$
$$= o(\max\{Q(q_s), Q(q_t)\}) ,$$

as $z \sim y$ and $w = o(y)$ and Q is monotonic. Then

$$(vT(q_v))^{-1} - (yT(q_y))^{-1} = O(Q(q_y)y^{-3}w) ,$$

and (5.27) yields

$$H(x,u) = H(x,y) - (u-y)^2(yT(q_y))^{-1} + O(Q(q_y)y^{-3}w^3) + O(w^2y^{-2})$$
$$= H(x,y) - (u-y)^2(yT(q_y))^{-1} + O(\{\log Q(q_y)\}^{3/2}Q(q_y)^{-1/2}) ,$$

by (5.24). Note that the order term is uniform in $u \in [y-w, y+w]$. Then, from (5.23) and (5.18),

(5.28) $$I(x) = \exp(H(x,y))(1+O(\{\log Q(x)\}^{3/2}Q(x)^{-1/2}))$$
$$\times \int_{y-w}^{y+w} \exp\{-(u-y)^2(yT(q_y))^{-1}\}du .$$

Here by (5.19),

(5.29) $$\exp(H(x,y)) = (2y/x)^j y^{-1/2}\exp(2Q(x))(1+O(1/Q(x)))$$
$$= (2Q'(x))^j y^{-1/2}\exp(2Q(x))(1+O(1/Q(x))) ,$$

by (5.22). Further, if $\chi := w(yT(q_y))^{-1/2}$,

(5.30) $$\int_{y-w}^{y+w} \exp(-(u-y)^2(yT(q_y))^{-1})du$$

$$= (yT(q_y))^{1/2} \int_{-\chi}^{\chi} \exp(-t^2)dt$$

$$= (yT(q_y))^{1/2}\, \pi^{1/2}(1+O(\chi^{-1}e^{-\chi^2})) .$$

Note that by (5.4), for some C independent of K,

$$(5.31) \qquad x \geq Cwy^{-1}Q(q_y)^{1/2} = CK(\log Q(q_y))^{1/2} .$$

Choosing K large enough, and combining (5.28) to (5.31), as well as (5.21), we obtain (5.26). □

Lemma 5.5

Assume the notation and assumptions of Lemma 5.2. Let $K > 0$ and y, w be as in Lemmas 5.3 and 5.4 respectively. Then if K is large enough, we have as $x \to \infty$,

$$(5.32) \qquad J(x) := \left(\int_1^{y-w} + \int_{y+w}^{\infty}\right)\exp(H(x,u))du$$

$$(5.33) \qquad = O(Q'(x)^j e^{2Q(x)} T(x)^{1/2} Q(x)^{-5}) .$$

Proof

We first estimate $\int_{y+w}^{\infty}$. Now, as in the proof of Lemma 5.3 (see also (5.13)), there exists $B > 0$ independent of x such that $H'(x,u)$ is strictly decreasing for u in (B,∞). Then as $H'(x,y) = 0$, and also $H'(x,u) < 0$, $u > y$,

$$(5.34) \qquad \int_{y+w}^{\infty} \exp(H(x,u))du \leq \int_{y+w}^{\infty} \exp(H(x,u))H'(x,u)/H'(x,y+w)du = -\exp(H(x,y+w))/H'(x,y+w) .$$

Next, there exists $v \in (w/2,w)$ such that

$$(5.35) \qquad H(x,y+w) = H(x,y+w/2) + (w/2)H'(x,y+v) \leq H(x,y) + (w/2)H'(x,y+w/2) ,$$

as both $H(x,u)$ and $H'(x,u)$ are decreasing for $u > y$. Let $a > 0$. Then as $w = o(y)$, (5.11) yields

$$H'(x,y+aw) = 2\log(x/q_y) + 2\log(q_y/q_{y+aw}) + O(y^{-1})$$

$$= -2\int_y^{y+aw} (uT(q_u))^{-1}du + O(y^{-1}) \qquad \text{(by (5.7), (5.15))}$$

$$\le - Caw\, y^{-2}Q(q_y) + O(y^{-1}) \qquad \text{(by (5.4))}$$

$$\le - CaK(Q(q_y)\log Q(q_y))^{1/2}y^{-1} + O(y^{-1}) ,$$

by (5.24). Combining this last inequality with (5.24), (5.34) and (5.35), and choosing $a = 1/2$ and $a = 1$,

$$\int_{y+w}^{\infty} \exp(H(x,u))du$$

$$\le C_2\exp(H(x,y) - C_1K(\log Q(q_y)))y\{Q(q_y)\log Q(q_y)\}^{-1/2} ,$$

where C_1, C_2 are independent of x and K. Combining (5.18), (5.19), (5.22) and this last inequality, we obtain

$$\int_{y+w}^{\infty} \exp(H(x,u))du$$

$$\le C_3Q'(x)^j\exp(2Q(x))Q(x)^{-C_1K-1/2}(xQ'(x))^{1/2}(\log Q(x))^{-1/2}$$

$$\le C_3Q'(x)^j\exp(2Q(x))Q(x)^{-C_1K}(T(x))^{1/2}(\log Q(x))^{-1/2} .$$

by (5.3). If $C_1K \ge 5$, this last right hand side is bounded above by the right hand side of (5.33). The estimation of $\int_1^{y-w}$ is similar: one uses the fact that for $y > u \ge B$, $H'(x,u)$ is positive and decreasing, while $H(x,u)$ is increasing. □

Proof of Theorem 3.1

By Lemmas 5.4 and 5.5,

$$(5.36)\qquad \int_1^\infty \exp(H(x,u))du = (2Q'(x))^j e^{2Q(x)}(\pi T(x))^{1/2} \times \{1 + O\{(\log Q(x))^{3/2}Q(x)^{-1/2}\}\}.$$

From (5.10), we see that it suffices to bound the series

$$\sum_{n>j/2} \exp(H(x,n))$$

in terms of the previous integral. But if $B + 1 < n < y - 1$,

$$\int_{n-1}^{n} \exp(H(x,u))du < \exp(H(x,n)) < \int_{n}^{n+1} \exp(H(x,u))du$$

while if $n > y + 1$,

$$\int_{n}^{n+1} \exp(H(x,u))du < \exp(H(x,n)) < \int_{n-1}^{n} \exp(H(x,u))du ,$$

by the monotonicity properties of $H(x,u)$. Since $H(x,y)$ is the maximum value of $H(x,u)$, and since the addition of any polynomial does not affect the asymptotic behaviour of $G_Q(x)$, we obtain as $x \to \infty$,

$$G_Q(x) = \int_1^\infty \exp(H(x,u))du + O(e^{H(x,y)}) .$$

Using (5.19), (5.22), (5.36) and (5.2), we obtain the result. □

Our proof of Theorem 3.2 will use:

Lemma 5.6

Let $W(x) := e^{-Q(x)}$ satisfy the hypotheses of Theorem 3.1, let j be a fixed non-negative integer, let $G_Q(x)$ be defined by (3.1), and let $S_{2n}(x)$ denote the (n+1)th partial sum of $G_Q(x)$. There exist C_1, C_2 and C_3 such that for n large enough and uniformly for $|x| \geq C_1$ satisfying

$$(5.37)\qquad xQ'(x)[1+C_2\{\log Q(x)/Q(x)\}^{1/2}] \leq n ,$$

we have

$$(5.38)\qquad |1-S_{2n}^{(j)}(x)/G_Q^{(j)}(x)| \le C_3 Q(x)^{-5} .$$

Proof

With the notation of Lemmas 5.2 and 5.4, for $x \ge C_1$,

$$|1-S_{2n}^{(j)}(x)/G_Q^{(j)}(x)| = \sum_{k=n+1}^{\infty} \exp(H(x,k))/G_Q^{(j)}(x) .$$

Here if $n \ge y + w$, this last right hand side is bounded above by

$$\int_{y+w}^{\infty} \exp(H(x,u))du/G_Q^{(j)}(x) = O(Q(x)^{-5}) ,$$

by (5.33) and Theorem 3.1. It remains to show that for C_2 large enough, (5.37) implies that $n \ge y + w$. Now by (5.18), (5.22) and (5.24),

$$\begin{aligned} y + w &= y(1+K\{\log Q(x)/Q(x)\}^{1/2}(1+o(1))) \\ &\le xQ'(x)(1+O(1/Q(x)))(1+C_4\{\log Q(x)/Q(x)\}^{1/2}) \\ &\le n , \end{aligned}$$

by (5.37) provided C_2 is large enough. □

Proof of Theorem 3.2

Setting $x = q_s$ in (2.10), where s is some large positive number, we obtain for each $\eta > 0$,

$$s/Q(q_s) = O(Q'(q_s)^{\eta}) = O((s/q_s)^{\eta}) .$$

It follows that given $\eta > 0$,

$$Q(q_s) \ge C_4 s^{1-\eta} , \quad \text{s large enough} .$$

Then if $x = q_s$, s large enough, and $\eta > 0$ is given,

$$f(x) := xQ'(x)[1+C_2\{\log Q(x)/Q(x)\}^{1/2}]$$

$$\leq s[1+C_3 s^{-1/2+\eta}] .$$

In particular, for

$$x = q_s \leq q_{n(1-n^{-1/2+\epsilon})} ,$$

$$f(x) \leq n(1-n^{-1/2+\epsilon})[1+C_3\{n(1-n^{-1/2+\epsilon})\}^{-1/2+\eta}]$$
$$\leq n ,$$

if we choose $\eta < \epsilon$. So (5.37) is satisfied for

$$C_5 \leq |x| \leq q_{n(1-n^{-1/2+\epsilon})} .$$

Then if $\{\xi_n\}_1^\infty$ is any sufficiently slowly increasing sequence of positive numbers with limit ∞, (5.38) yields

$$\lim_{n\to\infty} \|1 - S_{2n}^{(j)}(x)/G_Q^{(j)}(x)\|_{L_\infty(\xi_n\leq|x|\leq q_{n(1-n^{-1/2+\epsilon})})} = 0 .$$

Crude estimations of the tail of the series $G_Q^{(j)}(x)$ show that we can choose $\{\xi_n\}_1^\infty$ as above satisfying also

$$\lim_{n\to\infty} \|1 - S_{2n}^{(j)}(x)/G_Q^{(j)}(x)\|_{L_\infty(|x|\leq\xi_n)} = 0 .$$

For example, $\xi_n = q_n/2$ yields an easy proof of this last statement. Then the result follows. □

6. Proof of Theorem 3.3

In this section, we use the results of section 5, to prove Theorem 3.3. Throughout, we let

(6.1) $\Psi(x) := 1/\{G_Q(x)W^2(x)\}$, $x \in \mathbb{R}$.

One of the main lemmas of this section provides low degree polynomial approximations for $\Psi(x)$ in a suitable interval. First, some properties of $\Psi(x)$:

Lemma 6.1

Let $W := e^{-Q} \in$ VSE. Then for large enough x,

(6.2) $|\log \Psi(x)| = O(\log Q'(x))$,

and for each $\epsilon > 0$,

(6.3) $\frac{d}{dx} \log \Psi(x) = O(Q(x)^{1/2+\epsilon}(\log Q(x))^{3/2}x^{-1-\epsilon})$.

Proof

First note from (5.3) and (2.10) that for each $\epsilon > 0$,

(6.4) $T(x) = O(Q'(x)^{\epsilon})$, $x \to \infty$.

Further, (2.9) shows that

(6.5) $\lim_{x \to \infty} T(x) = +\infty$.

Now by Theorem 3.1, with $j = 0$, as $|x| \to \infty$,

$$\Psi(x) = (\pi T(x))^{-1/2}\{1+O(\{\log Q(x)\}^{3/2}Q(x)^{-1/2})\} .$$

Then as $|x| \to \infty$,

$$|\log \Psi(x)| \sim \log T(x)$$
$$= O(\log Q'(x)) ,$$

by (6.4). Hence (6.2).

Next, from (6.1), and Theorem 3.1 with $j = 0,1$,

$$\frac{d}{dx} \log \Psi(x) = - G_Q'(x)/G_Q(x) + 2Q'(x)$$

$$= -2Q'(x)\{1+O(\{\log Q(x)\}^{3/2}Q(x)^{-1/2})\} + 2Q'(x)$$
$$= O(Q'(x)\{\log Q(x)\}^{3/2}Q(x)^{-1/2}) .$$

Next, from (2.10), for each $\varepsilon > 0$, as $x \to \infty$,

(6.6) $$Q'(x)^{1-\varepsilon} = O(Q(x)/x) .$$

Hence given $\varepsilon > 0$, we have as $x \to \infty$,

(6.7) $$Q'(x) = O(\{Q(x)/x\}^{1+\varepsilon}) .$$

Then (6.3) follows. □

Lemma 6.2

Let $W := e^{-Q} \in$ VSE. Let $\varepsilon > 0$. Let $\{c_n\}_1^\infty$ be a sequence of positive numbers with

(6.8) $$\lim_{n \to \infty} c_n = \infty ,$$

and such that for some $0 < \Delta < 2$,

(6.9) $$Q'(c_n) = O(n^\Delta) , \quad n \to \infty .$$

Then there exists polynomials $U_n(x)$ of degree m_n, n large enough, with

(6.10) $$m_n = O(n^{\Delta/2+\varepsilon}) , \quad n \to \infty .$$

and

(6.11) $$\lim_{n \to \infty} \|U_n(x)\Psi^{-1}(c_n x)-1\|_{L_\infty[-1,1]} = 0 .$$

Proof

We shall write

$$\Psi = e^{\log \Psi} ,$$

and first approximate $\log \Psi$ by polynomials, then composing these approximations with the partial sums of the exponential function. First, from (2.9),

$$Q(c_n) = o(c_n Q'(c_n)) = o(c_n n^{\Delta}) , \quad n \to \infty .$$

Together with Lemma 4.2(i), (6.9) implies that for each $\delta > 0$,

$$c_n = O(n^{\delta}) , \quad n \to \infty .$$

Then

$$Q(c_n) = O(n^{\Delta+\delta}) , \quad n \to \infty ,$$

and

$$\log Q(c_n) = O(\log n) , \quad n \to \infty .$$

From (6.3), we deduce that given $\eta > 0$, we have for some C_1 and n large enough,

$$\max\{\frac{d}{dx} \log \Psi(c_n x) : C_1/c_n \leq |x| \leq 1\} \leq n^{\Delta/2+\eta} . \tag{6.12}$$

Now let $w_n(h)$ denote the usual modulus of continuity of the function $\log \Psi(c_n x)$ for $x \in [-1,1]$. Let $0 < \eta < a < \varepsilon$ and $\{\ell_n\}$ be a sequence of positive integers with

$$\ell_n \sim n^{\Delta/2+a} , \quad n \to \infty .$$

If u,v satisfy $C_1/c_n \leq u, v \leq 1$ and $|u-v| \leq 1/\ell_n$, then it follows from (6.12) that

$$\begin{aligned} |\log \Psi(c_n u) - \log \Psi(c_n v)| &\leq n^{\Delta/2+\eta}/\ell_n \\ &\to 0 , \quad n \to \infty , \end{aligned}$$

since $\eta < a$. If on the other hand $0 \leq u,v \leq C_1/c_n$, and $|u-v| \leq 1/\ell_n$,

$$|\log \Psi(c_n u) - \log \Psi(c_n v)|$$

$$\leq \max\{|\log \Psi(s) - \log \Psi(t)| : |s-t| \leq c_n/\ell_n ; |s| , |t| \leq C_1\}$$
$$\to 0 , \quad n \to \infty ,$$

since $c_n = o(\ell_n)$. These considerations show that

$$w_n(1/\ell_n) \to 0 , \quad n \to \infty .$$

By Jackson's Theorem, we can find a polynomial $R_n(u)$ of degree at most ℓ_n, such that

$$(6.13) \qquad \lim_{n \to \infty} \|\log \Psi(c_n u) - R_n(u)\|_{L_\infty[-1,1]} = 0 .$$

Next, let

$$\tau_k(u) := \sum_{j=0}^{k} u^j/j! ,$$

the (k+1)th partial sum of e^u. It is elementary, and well known that

$$(6.14) \qquad \tau_k(u) = e^u(1+o(1)) , \quad \text{uniformly for } |u| \le k/4 .$$

We note from (6.2) and (6.9) that

$$(6.15) \qquad \|\log \Psi(c_n u)\|_{L_\infty[-1,1]} = O(\log n) .$$

Assume, say, $\{k_n\}$ is a sequence of positive integers with

$$k_n \sim (\log n)^2 , \quad n \to \infty ,$$

and let

$$U_n(x) := \tau_{k_n}(R_n(x)) ,$$

a polynomial of degree at most

$$k_n \ell_n \sim (\log n)^2 \, n^{\Delta/2+a} = O(n^{\Delta/2+\epsilon}) .$$

Then from (6.13), (6.14) and (6.15), we have uniformly for $|x| \le 1$,

$$U_n(x)\Psi^{-1}(c_nx) = \exp(R_n(x)-\log \Psi(c_nx))(1+o(1))$$
$$= 1+o(1) \ . \qquad \square$$

Lemma 6.3

Let $W := e^{-Q} \in$ VSE and $g(x)$ be continuous in $\mathbb{R}$, with $g(x) = 0$ for $|x| \geq 1$. Let $\delta > 0$.

(i) There exists a sequence of positive numbers $\{c_n\}_1^\infty$ satisfying

(6.16) $$\lim_{n \to \infty} c_n/q_n = 1 \ ,$$

(6.17) $$\lim_{n \to \infty} Q(c_n)/n = \infty \ ,$$

and for each $\epsilon > 0$,

(6.18) $$Q'(c_n) = O(n^{1+\epsilon}) \ , \quad n \to \infty \ .$$

(ii) If $\{c_n\}_1^\infty$ is a sequence satisfying (6.16), (6.17) and (6.18), there exist for n large enough polynomials $\chi_n(u)$ of degree at most δn, such that

(6.19) $$\lim_{n \to \infty} \|g(x) - \chi_n(x)W^2(c_nx)\|_{L_\infty(\mathbb{R})} = 0 \ .$$

Proof

(i) We can choose $\{c_n\}$ satisfying (6.16) and (6.17), much as in the proof of Theorem 4.1. Further, noting by (4.13) that $Q(q_n) = o(n)$, we may choose $\{c_n\}$ to grow so that for n large enough,

$$Q(c_n) \leq n \log n \ .$$

Then (6.7) yields (6.18).

(ii) Choose polynomials $R_n(x)$ of degree at most log n, n large enough, such that

(6.20) $$\lim_{n\to\infty} \|g(u)-R_n(u)\|_{L_\infty[-2,2]} = 0 .$$

Let $U_n(x)$ be the polynomial of degree at most $n^{1/2+\epsilon}$, n large enough, such that (6.11) holds with some $\{c_n\}$ satisfying (6.16) to (6.18). Finally, let $S_{\delta n/2}(x)$ denote the at most $\delta n/4$-th partial sum of $G_Q(x)$, n large enough, and let

$$\chi_n(x) := R_n(x)U_n(x)S_{\delta n/4}(c_nx) ,$$

a polynomial of degree at most

$$\log n + n^{1/2+\epsilon} + \delta n/2 \le \delta n ,$$

n large enough. By Theorem 3.2, and by (6.11) and (6.20), we have if $|c_nx| \le q_{\delta n/8}$, n large enough,

$$\begin{aligned}&\chi_n(x)W^2(c_nx)\\ &= R_n(x)\{U_n(x)\Psi^{-1}(c_nx)\{S_{\delta n/4}(c_nx)/G_Q(c_nx)\}\\ &= (g(x) + o(1))(1+o(1))(1+o(1))\\ &= g(x) + o(1) ,\end{aligned}$$

and so

(6.21) $$\lim_{n\to\infty} \|\chi_n(x)W^2(c_nx) - g(x)\|_{L_\infty(|x|\le q_{\delta n/8}/c_n)} = 0 .$$

Next, since all terms of the series $G_Q(u)$ are non-negative,

$$0 \le S_m(u)/G_Q(u) \le 1 , \quad u \in \mathbb{R} .$$

Then (6.11), (6.20) and this last inequality show that for $q_{\delta n/8}/c_n \le |x| \le 1$,

$$\begin{aligned}&|\chi_n(x)W^2(c_nx) - g(x)|\\ &\le |R_n(x)|(1+o(1)) + |g(x)|\end{aligned}$$

$$\leq 2|g(x)| + o(1)$$
$$\to 0, \quad n \to \infty,$$

since $g(\pm 1) = 0$ and since (4.14) and (6.16) imply

$$\lim_{n \to \infty} q_{\delta n/8}/c_n = 1 .$$

Thus

$$(6.22) \qquad \lim_{n \to \infty} \| \chi_n(x) W^2(c_n x) - g(x) \|_{L_\infty[-1,1]} = 0 .$$

Since our choice of $\{c_n\}$ satisfies (4.7), we may apply (4.5) in Theorem 4.1 with $\hat{W} := W^2$ and $p := \infty$ to deduce that

$$\begin{aligned} \| \chi_n(x) W^2(c_n x) \|_{L_\infty(|x| \geq 1)} &= \| \chi_n(x/c_n) W^2(x) \|_{L_\infty(|x| \geq c_n)} \\ &\leq \delta_n \| \chi_n(x/c_n) W^2(x) \|_{L_\infty(|x| \leq c_n)} \\ &= \delta_n \| \chi_n(x) W^2(c_n x) \|_{L_\infty[-1,1]} \\ &\to 0, \quad n \to \infty, \end{aligned}$$

by (4.4) and (6.22). Since $g(x) = 0$ for $|x| \geq 1$, (6.19) then follows. □

Proof of Theorem 3.3

We first show (ii) ⇒ (i). So let $g(x)$ be continuous in $\mathbb{R}$ with $g(x) = 0$ for $|x| \geq 1$. We shall apply Lemma 6.3 but with $W := e^{-Q}$ replaced by $W^{1/2} = e^{-Q/2}$, which also belongs to VSE. Note that working with $Q/2$ rather than Q involves only replacement of q_n by q_{2n}. Since by Lemma 4.2,

$$q_{2n}/q_n \to 1, \quad n \to \infty,$$

Lemma 6.3 shows that if $\{c_n\}_1^\infty$ satisfies (6.16) to (6.18), we can find $\chi_n(x)$ of degree at most δn with

(6.23) $$\lim_{n\to\infty} \|g(u) - \chi_n(u)W(c_n u)\|_{L_\infty(\mathbb{R})} = 0 .$$

The case of general $\{c_n\}$ satisfying only (3.6) may be achieved using a substitution: Let $\{d_n\}$ satisfy

$$\lim_{n\to\infty} d_n/Q^{[-1]}(n) = 1 .$$

Then also

$$\lim_{n\to\infty} d_n/c_n = 1 ,$$

by Lemma 4.2 (vii) and (6.16). Making the substitution $c_n u = d_n x$ in (6.23) and letting $\chi_n(u) =: Z_n(x)$, we obtain

$$\lim_{n\to\infty} \|g(d_n c_n^{-1} x) - Z_n(x)W(d_n x)\|_{L_\infty(\mathbb{R})} = 0 .$$

Since g is continuous in $\mathbb{R}$ and vanishes outside $(-1,1)$, we have

$$\lim_{n\to\infty} \|g(x) - g(d_n c_n^{-1} x)\|_{L_\infty(\mathbb{R})} = 0 .$$

This establishes (ii) $\Rightarrow$ (i) in general.

To show (i) $\Rightarrow$ (ii), we note that we may assume the given $\{c_n\}$ satisfies (4.7) in Theorem 4.1 - otherwise make a substitution as above. Then (4.5) in Theorem 4.1 with $\hat{W} := W$, $p = \infty$, and (3.7) yield

$$\lim_{n\to\infty} \|P_n(x)W(c_n x)\|_{L_\infty(|x|\geq r)}$$

$$= \lim_{n\to\infty} \|P_n(x/c_n)W(x)\|_{L_\infty(|x|\geq rc_n)} = 0 ,$$

for each $r > 1$. Then (3.7) yields $g(x) = 0$ in $|x| \geq r$, for each $r > 1$. □

7. Proof of Theorems 2.3, 2.4 and 2.5

Our proof will make use of a result obtained jointly by the authors and P. Nevai [10]:

Lemma 7.1

Let $\hat{W}^2$ *be a weight function. Suppose there exists an increasing sequence of positive numbers* $\{c_n\}_1^\infty$ *and a decreasing sequence of positive numbers* $\{\delta_n\}_1^\infty$ *such that*

$$(7.1) \qquad \lim_{n \to \infty} \delta_n = 0 \ ,$$

and such that for $n = 1,2,3,\ldots$ *and* P *of degree* $\leq n$,

$$(7.2) \qquad \|P\hat{W}\|_{L_2(\mathbb{R})} \leq (1+\delta_n)\|P\hat{W}\|_{L_2(-c_n,c_n)} \ .$$

Suppose further that there exist real polynomials $R_{n-2}(x)$ *of degree at most* $n-2$: $n = 2,3,4,\ldots$, *such that for* $p = 1/2$ *and* $p = 2$,

$$(7.3) \qquad \lim_{n \to \infty} \pi^{-1} \int_{-1}^{1} |\hat{W}(c_n x)R_{n-2}(x)(1-x^2)^{1/4}|^p dx/\sqrt{1-x^2} = 1 \ .$$

Then

$$(7.4) \qquad \lim_{n \to \infty} A_n(\hat{W}^2)/c_n = 1/2 \ .$$

and

$$(7.5) \qquad \lim_{n \to \infty} B_n(\hat{W}^2)/c_{n+1} = 0 \ .$$

Proof of Theorem 2.5

Note first from (2.21) and the other hypotheses on $\hat{W}$ that

$$\lim_{|x| \to \infty} \{\log 1/\hat{W}(x)\}/Q(x) = 1 \ .$$

Hence Theorem 4.1 shows that (7.2) is true with the given choice of

$\{c_n\}$, since (2.22) and (2.23) are satisfied. Next, if $g(x)$ is continuous in $\mathbb{R}$ with $g(x) = 0$, $|x| > 1$, Theorem 3.3 shows that there exist polynomials $P_n(x)$ of degree $\leq n/4$ such that

$$(7.6) \qquad \lim_{n\to\infty} \|g(x) - P_n(x)W(c_nx)\|_{L_\infty(\mathbb{R})} = 0 .$$

Let $U_n(x)$ be the polynomial of degree $\leq n/2$ in Theorem 2.5 satisfying (2.24) and (2.25).

To deal with the factor $w_F(x)$, we proceed in a slightly different way from that in [15]. With the notation of Definition 2.2, let

$$\Delta := \sum_{j=1}^{N} \Delta_j ,$$

and ℓ be a positive integer such that

$$\ell > \max_j |\Delta_j| .$$

Let

$$Y(x) := \prod_{j=1}^{N} (x-z_j)^{\ell} .$$

We see that for some $C > 0$ and all n large enough,

$$(7.7) \qquad |w_F(c_nx)Y(c_nx)|c_n^{-\Delta-N\ell} \leq C , \quad x \in [-1,1] ,$$

and

$$(7.8) \qquad \lim_{n\to\infty} |w_F(c_nx)Y(c_nx)|c_n^{-\Delta-N\ell} = |x|^{\Delta+N\ell} , \quad x \neq 0 .$$

Now let

$$\hat{R}_n(x) := P_n(x)U_n(x)Y(c_nx)c_n^{-\Delta-N\ell} ,$$

a polynomial of degree at most $n/4 + n/2 + N\ell < 7/8n$, n large enough. We have from (7.6), (7.7), (2.25) and (2.26) that

$$|\hat{R}_n(x)W(c_nx)| = \varphi(c_nx)|P_n(x)W(c_nx)|\ |V(c_nx)U_n(x)|$$
$$\times\ |w_F(c_nx)Y(c_nx)c_n^{-\Delta-N\ell}|$$
$$\leq C, \quad x \in [-1,1].$$

Further, by (2.24), (7.6) and (7.8), we have for a.e. $x \in [-1,1]$,

$$\lim_{n\to\infty} |\hat{R}_n(x)W(c_nx)| = g(x)|x|^{\Delta+N\ell}.$$

Hence for $p = 1/2, 2$,

$$\lim_{n\to\infty} \pi^{-1}\int_{-1}^{1} |\hat{W}(c_nx)\hat{R}_n(x)(1-x^2)^{1/4}|^p dx/\sqrt{1-x^2}$$

$$= \pi^{-1}\int_{-1}^{1} |g(x)|x|^{\Delta+N\ell}(1-x^2)^{1/4}|^p dx/\sqrt{1-x^2}.$$

But we can choose $g(x)$ such that

$$0 \leq g(x)|x|^{\Delta+N\ell}(1-x^2)^{1/4} \leq 1, \quad x \in (-1,1),$$

with equality outside a set of arbitrarily small measure. It follows that we can choose $R_n(x)$ of degree $\leq 7n/8$, n large enough, such that for $p = 1/2, 2$,

$$\lim_{n\to\infty} \pi^{-1}\int_{-1}^{1} |\hat{W}(c_nx)R_n(x)(1-x^2)^{1/4}|^p dx/\sqrt{1-x^2}$$

$$= \pi^{-1}\int_{-1}^{1} dx/\sqrt{1-x^2} = 1.$$

Hence (7.3) holds and so (7.4) and (7.5) are applicable. □

Proof of Theorem 2.4

Fix $\{c_n\}$ satisfying (4.3) and (4.7). Since $Q(q_n) = o(n)$, we may assume that for n large enough,

$$Q(c_n) \leq n \log n,$$

so that

$$c_n \leq Q^{[-1]}(n \log n) .$$

Let η be a small positive number, and for n large enough, let

$$r_n := Q^{[-1]}(n^{1+\eta}) .$$

Let $\epsilon > 0$. We have

$$\text{(7.9)} \qquad c_n/r_n \leq Q^{[-1]}(n \log n)/Q^{[-1]}(n^{1+\eta})$$

$$= \exp(-\int_{n \log n}^{n^{1+\eta}} \{Q^{[-1]}(u)Q'(Q^{[-1]}(u))\}^{-1}du)$$

$$\leq \exp(-\int_{n \log n}^{n^{1+\eta}} u^{-1-\epsilon}du) \leq \exp(-Cn^{-\epsilon(1+\eta)}) ,$$

where we have used (6.7). Next, let $U_n(x)$ denote the at most n/2-th partial sum of the Maclaurin series of the entire function $e^{-g(x)}$. Using the usual contour integral error formula for $U_n(x) - e^{-g(x)}$, we obtain

$$\|U_n(x)e^{g(x)}-1\|_{L_\infty(-c_n,c_n)}$$

$$\leq \exp(2\max\{|g(t)| : |t| \leq r_n\}) (c_n/r_n)^{n/2}(1-c_n/r_n)^{-1}$$

$$\leq \exp(O(Q(r_n)^{1-\delta}) - Cn^{1-\epsilon(1+\eta)}/3) \quad \text{(by (2.20))}$$

$$\rightarrow 0 , \quad n \rightarrow \infty ,$$

by choice of r_n, if ϵ, η are small enough. Then (2.24) and (2.25) hold for $V := e^g$ and Theorem 2.5 yields the result. □

Proof of Theorem 2.3

We show that $V := e^g$ satisfies the hypotheses (2.21), (2.24) and (2.25) of Theorem 2.5. Choose $\{c_n\}$ to satisfy (2.22), (2.23) and

$$Q(c_n) \le n(\log q_n)^{1/2} ,$$

n large enough. By (2.15), and since $c_n/q_n \to 1$,

$$\|g'\|_{L_\infty[-c_n,c_n]} = O(n^{1/2}(\log q_n)^{-3/4}/q_n) .$$

Then if $\{m_n\}$ are positive integers with

$$m_n \sim n^{1/2}(\log q_n)^{-1/2} ,$$

Jackson's theorem shows that we can find polynomials $P_n(u)$ of degree at most m_n such that

$$\lim_{n\to\infty} \|g(c_n x) - P_n(x)\|_{L_\infty[-1,1]} = 0 .$$

Note too that

$$\|g\|_{L_\infty[-c_n,c_n]} = O(c_n\|g'\|_{L_\infty[-c_n,c_n]} + 1)$$

$$= O(n^{1/2}(\log q_n)^{-3/4}) .$$

If $\tau_k(u)$ denotes the (k+1)th partial sum of e^u, we obtain much as in the proof of Lemma 6.2 that

$$U_n(x) := \tau_{k_n}(-P_n(x))$$

satisfies

$$\lim_{n\to\infty} \|U_n(x)e^{g(x)}-1\|_{L_\infty(-c_n,c_n)} = 0 ,$$

provided

$$k_n \ge 8\|g\|_{L_\infty[-c_n,c_n]} .$$

We can choose

$$k_n \sim n^{1/2}(\log q_n)^{-3/4} .$$

Then $U_n(x)$ has degree at most

$$k_n m_n \sim n(\log q_n)^{-5/4} = o(n) . \qquad \square$$

Acknowledgement. The authors would like to thank A.L. Levin, P. Nevai and E.B. Saff for encouragement and references, and would also like to thank Mrs L. Lochtenbergh for typing this long manuscript.

References

1. W.C. Bauldry, A. Máté and P. Nevai, Asymptotics for Solutions of Systems of Smooth Recurrence Equations, to appear in Pacific J. Math.
2. Z. Ditzian and V. Totik, Moduli of Smoothness, to appear.
3. P. Erdös, On the Distribution of the Roots of Orthogonal Polynomials, (in) Proc. Conf. Constr. Th. Fns. (eds. G. Alexits, et al.), Adademiai Kiado, Budapest, 1972, pp. 145-150.
4. G. Freud, Orthogonal Polynomials, Akademiai-Kiado, Pergamon Press, Budapest, 1971.
5. G. Freud, On the Coefficients in the Recursion Formulae of Orthogonal Polynomials, Proc. Roy. Irish. Acad., Sect. A., 76 (1976), 1-6.
6. G. Freud, On Markov-Bernstein Type Inequalities and Their Applications, J. Approx. Th., 19 (1977), 22-37.
7. G. Freud, On the Greatest Zero of an Orthogonal Polynomial, J. Approx. Th., 46 (1986), 16-24.
8. J. Geronimo and W. Van Assche, Orthogonal Polynomials with Asymptotically Periodic Recurrence Coefficients, J. Approx. Th., 46 (1986), 251-284.
9. A.A. Goncar and E. Rahmanov, Equilibrium Measure and the Distribution of Zeros of Extremal Polynomials, Math. USSR, Sbornik, 53 (1986), 119-130.
10. A. Knopfmacher, D.S. Lubinsky and P. Nevai, Freud's Conjecture and Approximation of Reciprocals of Weights by Polynomials, manuscript.
11. D.S. Lubinsky, A Weighted Polynomial Inequality, Proc. of the Amer. Math. Soc., 92 (1984), 263-267.
12. D.S. Lubinsky, Gaussian Quadrature, Weights on the Whole Real Line and Even Entire Functions with Nonnegative Even Order

Derivatives, J. Approx. Th., 46 (1986), 297-313.

13. D.S. Lubinsky, Even Entire Functions Absolutely Monotone in $[0,\infty)$ and Weights on the Whole Real Line, (in) Orthogonal Polynomials and Their Applications (C. Brezinski, et al., eds.), Springer Lecture Notes in Mathematics, Vol. 1171, Berlin, 1986.
14. D.S. Lubinsky, H.N. Mhaskar and E.B. Saff, Freud's Conjecture for Exponential Weights, Bull. of the Amer. Math. Soc., 15 (1986), 217-221.
15. D.S. Lubinsky, H.N. Mhaskar and E.B. Saff, A Proof of Freud's Conjecture for Exponential Weights, to appear in Constructive Approximation.
16. D.S. Lubinsky and E.B. Saff, Uniform and Mean Approximation by Certain Weighted Polynomials, with Applications, Constr. Approx 4 (1988).
17. D.S. Lubinsky and E.B. Saff, Strong Asymptotics for Extremal Errors and Extremal Polynomials Associated with Weights on $(-\infty,\infty)$, to appear in Springer Lecture Notes in Mathematics.
18. Al. Magnus, A Proof of Freud's Conjecture About Orthogonal Polynomials Related to $|x|^{\rho}\exp(-x^{2m})$, (in) Orthogonal Polynomials and Their Applications (C. Brezinski, et al., eds.), Springer Lecture Notes in Mathematics, Vol. 1171, Berlin, 1986.
19. Al. Magnus, On Freud's Conjecture for Exponential Weights, J. Approx. Th., 46 (1986), 65-99.
20. A. Máté and P. Nevai, Asymptotics for Solutions of Smooth Recurrence Relations, Proc. of the Amer. Math. Soc., 93 (1985), 423-429.
21. A. Máté, P. Nevai and T. Zaslavsky, Asymptotic Expansion of Ratios of Coefficients of Orthogonal Polynomials with Exponential Weights, Trans. of the Amer. Math. Soc., 287 (1985), 495-505.
22. H.N. Mhaskar and E.B. Saff, Extremal Problems for Polynomials with Exponential Weights, Trans. of the Amer. Math. Soc., 285 (1984), 203-234.
23. H.N. Mhaskar and E.B. Saff, Weighted Polynomials on Finite and Infinite Intervals: A Unified Approach, Bull. of the Amer. Math. Soc., 11 (1984), 351-354.
24. H.N. Mhaskar and E.B. Saff, Where Does the Sup Norm of a Weighted Polynomial Live? (A Generalization of Incomplete Polynomials), Constr. Approx., 1 (1985), 71-91.
25. H.N. Mhaskar and E.B. Saff, Where Does the L_p Norm of a Weighted Polynomial Live?, to appear in Trans. of the Amer. Math. Soc.
26. P. Nevai, Orthogonal Polynomials, Memoirs of the Amer. Math. Soc., 213 (1979), 1-85.

27. P. Nevai, Asymptotics for Orthogonal Polynomials Associated with $\exp(-x^4)$, SIAM J. Math. Anal., 15 (1984), 1177-1187.
28. P. Nevai, G. Freud, Christoffel Functions and Orthogonal Polynomials, A Case Study, J.Approx. Theory, 48(1986), 3-167.
29. E.A. Rahmanov, On Asymptotic Properties of Polynomials Orthogonal on the Real Axis, Math. USSR. Sbornik, 47 (1984), 155-193.
30. E.A. Rahmanov, Dissertation, Moscow, 1983.
31. E.B. Saff, Incomplete and Orthogonal Polynomials, (in) Approximation Theory IV, (eds. C.K. Chui, et al.), New York, Academic Press, 1983, pp. 219-256.
32. W. Van Assche, Asymptotic Properties of Orthogonal Polynomials from their Recurrence Formula II, to appear in J. Approx. Th.
33. W. Van Assche and J. Geronimo, Asymptotics for Orthogonal Polynomials with Unbounded Recurrence Coefficients, manuscript.

SOME EXAMPLES IN APPROXIMATION ON THE UNIT DISK BY RECIPROCALS OF POLYNOMIALS

by

A.L. Levin* and E.B. Saff**

Department of Mathematics
Everyman's University
16 Klausner St. P.O.B. 39328
Tel-Aviv 61392
ISRAEL

Institute for Constructive Mathematics
Department of Mathematics
University of South Florida
Tampa, FL 33620

Abstract. This paper is a continuation of the authors' study of approximation by reciprocals of polynomials. A Jackson-type theorem for such approximants is established for a certain class of functions f analytic and nonzero in the disk $|z| < 1$ and continuous on $|z| \leq 1$. Furthermore, we obtain the sharp degree of convergence for reciprocal polynomial approximation on $|z| \leq 1$ to functions f that are analytic on $|z| \leq 1$, nonzero in $|z| < 1$, and vanish somewhere on $|z| = 1$.

1. Statement of results.

In our papers [1], [2] we investigated the rate of approximation of real and complex-valued functions on $[-1,1]$ by reciprocals of polynomials. Here we extend some of these results to the case of

---

*The research of this author was conducted while visiting the Institute for Constructive Mathematics at the University of South Florida.

**The research of this author was supported, in part, by the National Science Foundation.

AMS subject classification 41A20, 41A17

approximation on the unit disk of the complex plane $\mathbb{C}$.

For any continuous function f in the closed unit disk $D = \{z: |z| \le 1\}$, let $E_{on}(f;D)$ denote the error in best uniform approximation of f on D by reciprocals of polynomials of degree $\le n$. J.L. Walsh [4] proved that $E_{on}(f;D) \to 0$ as $n \to \infty$ if and only if the continuous function f is analytic in the open disk $|z| < 1$ and does not vanish there. We denote the set of all such functions by $A_o(D)$. Under an additional assumption on f we can prove the following analogue of Jackson's theorem:

<u>Theorem</u> 1. <u>Let</u> $f \in A_o(D)$ <u>and suppose that the set</u> $\{f(z): z \in D\}$ <u>lies in a half-plane</u> $Re(z\overline{z}_o) \ge 0$, <u>for some</u> z_o, $|z_o| = 1$. <u>Then there exists a constant</u> M <u>(independent of</u> f <u>and</u> z_o<u>) such that</u>

(1.1) $$E_{on}(f;D) \le M\omega(f;n^{-1}), \qquad n = 1,2,3,\ldots ,$$

<u>where</u> $\omega(f;\delta)$ <u>denotes the modulus of continuity of</u> $f(e^{i\theta})$ <u>on</u> $[-\pi,\pi]$.

<u>Example</u>. It is easy to see that any single-valued branch of the function $(1-z)^\alpha$ satisfies the assumptions of Theorem 1 provided $0 < \alpha \le 1$. It follows that there exists a constant c such that

(1.2) $$E_{on}((1-z)^\alpha;D) \le cn^{-\alpha}, \quad 0 < \alpha \le 1; \quad n = 1,2,3,\ldots .$$

It can be also shown that the estimate (1.2) is precise in the sense that there exists a constant $d > 0$ such that

(1.3) $$E_{on}((1-z)^\alpha;D) \ge dn^{-\alpha}, \quad 0 < \alpha \le 1, \quad n = 1,2,3,\ldots .$$

From (1.3) it follows that the estimate given in Theorem 1 is, in general, the best possible.

The asymptotic character of $E_{on}(f;D)$ can be described precisely if we assume that f is analytic in the <u>closed</u> unit disk.

<u>Theorem</u> 2. <u>Let</u> $f \in A_o(D)$ <u>be analytic in the closed unit disk</u> D <u>and assume</u> f <u>vanishes somewhere on</u> $|z| = 1$. <u>Denote by</u> r <u>the smallest order of zeros of</u> f <u>on</u> $|z| = 1$. <u>Then there exist positive</u>

<u>constants</u> $A(f)$, $B(f)$ <u>such that</u>

$$A(f)n^{-r} \le E_{on}(f;D) \le B(f)n^{-r}, \qquad n = 1,2,3,\ldots . \tag{1.4}$$

<u>In particular, for any positive integer</u> r <u>there exist positive constants</u> A_r, B_r <u>such that</u>

$$A_r n^{-r} \le E_{on}((1-z)^r;D) \le B_r n^{-r}, \qquad n = 1,2,3,\ldots . \tag{1.5}$$

Finally, we mention the result of Walsh [4, Theorem V] that describes completely the functions for which $E_{on}(f;D)$ decreases exponentially.

<u>Theorem</u> 3 (Walsh). <u>For any continuous function</u> $f(\not\equiv 0)$ <u>on</u> D <u>the following conditions are equivalent</u>:

(i) $\limsup_{n\to\infty}[E_{on}(f;D)]^{1/n} \le 1/R < 1$.

(ii) f <u>is analytic on</u> D <u>and meromorphic and different from zero in</u> $D_R := \{z: |z| < R\}$.

2. Proof of Theorem 1.

By the assumption on f there exists z_0, $|z_0| = 1$, such that

$$\mathrm{Re}(f(z)\bar{z}_0) \ge 0, \qquad z \in D.$$

Consider the function

$$G(z) := f(z) + Az_0\omega(f;n^{-1}), \tag{2.1}$$

where $A > 0$ will be chosen later. Notice that

$$|G(z)| = |G(z)\bar{z}_0| = |f(z)\bar{z}_0 + A\omega(f;n^{-1})| \ge A\omega(f;n^{-1}), \; z \in D. \tag{2.2}$$

Now set

$$g(\theta) := G(e^{i\theta}), \qquad -\pi \le \theta \le \pi.$$

From (2.2) it follows that

$$|g(\theta)| \geq A\omega(f;n^{-1}), \qquad -\pi \leq \theta \leq \pi. \tag{2.3}$$

Furthermore,

$$\omega(g;n^{-1}) = \omega(f;n^{-1}) \tag{2.4}$$

(recall that $\omega(f;n^{-1})$ denotes the modulus of continuity of the function $f(e^{i\theta})$ on $[-\pi,\pi]$).

Let $K_n(t)$ be the Jackson kernel (see Lorentz [3, p.55]). Since

$$\int_{-\pi}^{\pi} K_n(t)dt = 1, \qquad \int_{-\pi}^{\pi} |t^k|K_n(t)dt = O(n^{-k}), \qquad k = 1,2,$$

we obtain for all θ that

$$\int_{-\pi}^{\pi} |g(\theta+t) - g(\theta)|^k K_n(t)dt \leq c[\omega(g;n^{-1})]^k, \qquad k = 1,2, \tag{2.5}$$

where $c > 0$ is an absolute constant. Now define

$$p_n(\theta) := \int_{-\pi}^{\pi} \frac{1}{g(\theta+t)} K_n(t)dt. \tag{2.6}$$

It is well-known that $p_n(\theta)$ has the form $\sum_{k=-n}^{n} \lambda_k c_k e^{ik\theta}$, where $\sum_{-\infty}^{\infty} c_k e^{ik\theta}$ is the Fourier series of $1/g(\theta)$. Since $G \neq 0$ in D (by (2.2)), $1/G$ is analytic in $|z| < 1$ and consequently $c_k = 0$ for $k < 0$. It follows that $p_n(\theta)$ is a polynomial in $e^{i\theta}$ of degree $\leq n$. We shall use the notation $P_n(z)$ for the corresponding algebraic polynomial in z, that is, $P_n(z) = \sum_{0}^{n} \lambda_k c_k z^k$. Now,

$$\left|\frac{1}{g(\theta)} - p_n(\theta)\right| = \left|\int_{-\pi}^{\pi} [1/g(\theta) - 1/g(\theta+t)]K_n(t)dt\right|$$

$$\leq \int_{-\pi}^{\pi} \frac{|g(\theta+t) - g(\theta)|}{|g(\theta)||g(\theta+t)|} K_n(t)dt$$

$$\leq \frac{1}{|g(\theta)|A\omega(f;n^{-1})} \int_{-\pi}^{\pi} |g(\theta+t) - g(\theta)|K_n(t)dt \qquad \text{(by (2.3))}$$

$$\leq \frac{1}{|g(\theta)|A\omega(f;n^{-1})} c\omega(f;n^{-1}) \qquad \text{(by (2.5),(2.4))}$$

$$= \frac{c}{|g(\theta)|A} .$$

The choice

(2.7) $\quad A := 2c$

therefore yields

(2.8) $\quad |1-g(\theta)p_n(\theta)| \leq 1/2 \quad , \quad -\pi \leq \theta \leq \pi,$

which implies that

(2.9) $\quad |g(\theta)p_n(\theta)| \geq 1/2 \quad , \quad -\pi \leq \theta \leq \pi.$

From (2.8) we deduce, by the maximum principle, that $|1-G(z)P_n(z)| \leq 1/2$ for $|z| \leq 1$ and therefore

$$|G(z)P_n(z)| \geq 1/2 \quad , \quad |z| \leq 1.$$

In particular, $P_n(z) \neq 0$ in D and applying the maximum principle again we conclude that

(2.10) $$\max_{|z|\leq 1} |G(z) - 1/P_n(z)| = \max_{-\pi\leq\theta\leq\pi} |g(\theta) - 1/p_n(\theta)|.$$

Now,

$$|g(\theta) - 1/p_n(\theta)| = \left|\int_{-\pi}^{\pi} \frac{g(\theta+t) - g(\theta)}{g(\theta)g(\theta+t)} \cdot \frac{g(\theta)}{p_n(\theta)} \cdot K_n(t)dt\right|$$

$$\leq 2\int_{-\pi}^{\pi} |g(\theta+t) - g(\theta)|\cdot\left|\frac{g(\theta)}{g(\theta+t)}\right|\cdot K_n(t)dt \quad \text{(by (2.9))}$$

$$\leq 2\int_{-\pi}^{\pi} |g(\theta+t)-g(\theta)|K_n(t)dt + 2\int_{-\pi}^{\pi} \frac{|g(\theta+t)-g(\theta)|^2}{|g(\theta+t)|} K_n(t)dt$$

$$\leq 2c\omega(f;n^{-1}) + \frac{2}{2c\omega(f;n^{-1})} c[\omega(f;n^{-1})]^2$$

$$\text{(by (2.5), (2.4) and (2.7))}$$

$$= (2c + 1)\omega(f;n^{-1}).$$

From (2.10) and from the definition (2.1) of G it now follows (see also (2.7)) that

$$\max_{|z|\leq 1} |f(z) - 1/P_n(z)| \leq (4c + 1)\omega(f;n^{-1}). \qquad \square$$

3. Proof of Theorem 2.

To establish the upper bound in (1.4), we first prove that, for each positive integer r,

$$E_{on}((1-z)^r;D) \leq B_r n^{-r}, \qquad n = 1,2,3,\ldots . \tag{3.1}$$

Define

$$p(z) := \left[\frac{1 - Q(z)^r}{1-z}\right]^r, \qquad n \geq 2, \tag{3.2}$$

where

$$Q(z) := \frac{1-z^n}{n(1-z)} .$$

Since

$$Q(z) = \frac{1-z^n}{n(1-z)} = 1 - \frac{n-1}{2}(1-z) + O((1-z)^2) ,$$

$p(z)$ is a polynomial (of degree $(n-1)r^2 - r$) satisfying $p(1) = (r(n-1)/2)^r$. Also,

$$(3.3)\qquad |Q(z)| = \frac{1}{n}\,|1+z+\cdots+z^{n-1}| < 1 \quad \text{for} \quad |z| \le 1,\ z \ne 1.$$

It follows that $p(z) \ne 0$ in D and consequently it suffices to estimate $|(1-z)^r - 1/p(z)|$ on $|z| = 1$. Since $p(z)$ has real coefficients we may restrict ourselves to the case $z = e^{i\theta}$, $0 \le \theta \le \pi$.

Case 1. $\pi^2/2n \le \theta \le \pi$.
In this case

$$n|1-z| = 2n\sin(\theta/2) \ge 2n\sin(\pi^2/4n) \ge 2n\cdot\frac{2}{\pi}\cdot\frac{\pi^2}{4n} = \pi,$$

so that

$$(3.4)\qquad |Q(z)| = \left|\frac{1-z^n}{n(1-z)}\right| \le \frac{2}{\pi}.$$

Now write

$$(3.5)\qquad (1-z)^r - \frac{1}{p(z)} = (1-z)^r\,\frac{[(1-Q(z)^r)^r - 1]}{[1-Q(z)^r]^r}$$

$$= \frac{-Q(z)^r(1-z)^r}{[1-Q(z)^r]^r}\sum_{k=0}^{r-1}[1-Q(z)^r]^k.$$

Using (3.3), (3.4) and the obvious inequality $|Q(z)(1-z)| \le 2/n$, we obtain

$$|(1-z)^r - \frac{1}{p(z)}| \le \frac{2^r n^{-r}\sum_{k=0}^{r-1}2^k}{(1-(2/\pi)^r)^r} =: c_r n^{-r},$$

where c_r depends only on r.

Case 2. $2\varepsilon/n \le \theta < \pi^2/2n$, for some $0 < \varepsilon < 1$.
In this case $\theta < 2\pi/n$ and since the function $(\sin(n\theta/2))/\sin(\theta/2)$ is decreasing for $0 < \theta < 2\pi/n$, we obtain

$$|Q(z)| = \left|\frac{1-z^n}{n(1-z)}\right| = \left|\frac{\sin(n\theta/2)}{n\sin(\theta/2)}\right| \le \frac{\sin\varepsilon}{n\sin(\varepsilon/n)}.$$

Using the Maclaurin development for the sine function one can easily show that

$$\frac{\sin \varepsilon}{n \sin(\varepsilon/n)} < 1 - \varepsilon^2/10 \quad \text{for} \quad 0 < \varepsilon < 1, \quad n \geq 2,$$

and therefore

$$|Q(z)|^r \leq (1 - \varepsilon^2/10)^r \leq 1 - \varepsilon^2/10$$

which implies

$$|1 - Q(z)^r| \geq \varepsilon^2/10 \quad \text{for} \quad 0 < \varepsilon < 1, \quad n \geq 2.$$

Using this estimate together with (3.3) and $|Q(z)(1-z)| \leq 2/n$ we obtain from (3.5) that

$$\left|(1-z)^r - \frac{1}{p(z)}\right| \leq \frac{2^r n^{-r} \sum_{k=0}^{r-1} 2^k}{(\varepsilon^2/10)^r} =: c_r \varepsilon^{-2r} n^{-r} ,$$

where c_r depends only on r.

<u>Case 3</u>. $0 < \theta < 2\varepsilon/n$, $\varepsilon > 0$ is small enough.

In this case

(3.6) $$|1-z| < 2\varepsilon/n$$

so that

(3.7) $$|1-z|^r < (2\varepsilon/n)^r < n^{-r} \quad \text{if} \quad \varepsilon < 1/2.$$

Next, we write $p(z)$ in (3.2) in the form

(3.8) $$p(z) = \left[\frac{1 - Q(z)}{1-z}\right]^r \cdot \left[\sum_{k=0}^{r-1} Q(z)^k\right]^r ,$$

where

$$Q(z) = \frac{1-z^n}{n(1-z)} = 1 + \frac{1}{n} \sum_{j=2}^{n} \binom{n}{j} (z-1)^{j-1} .$$

Since

$$\left|\frac{1}{n}\sum_{j=2}^{n}\binom{n}{j}(z-1)^{j-1}\right| \le \sum_{j=2}^{n}\frac{1}{j!}(2\epsilon)^{j-1} \qquad \text{(by (3.6))}$$

$$= 2\epsilon\sum_{j=2}^{n}\frac{1}{j!}(2\epsilon)^{j-2}$$

$$\le 2\epsilon e \quad , \quad \text{if} \quad \epsilon < 1/2,$$

we obtain

$$\sum_{k=0}^{r-1} Q(z)^k = \begin{cases} 1 & \text{if} \quad r = 1 \\ r + O(\epsilon) & \text{if} \quad r \ge 2, \end{cases}$$

where $O(\epsilon)$ depends only on r. It follows that there exists ϵ_r, $0 < \epsilon_r < 1/2$ (that depends only on r), such that

$$(3.9) \qquad \left|\sum_{k=0}^{r-1} Q(z)^k\right| \ge 1 \quad , \quad \text{provided} \quad 0 < \theta < 2\epsilon_r/n.$$

From (3.7), (3.8), and (3.9) we obtain for $z = e^{i\theta}$, $0 < \theta < 2\epsilon_r/n$,

$$(3.10) \qquad \left|(1-z)^r - 1/p(z)\right| \le |1-z|^r + |1/p(z)|$$

$$\le n^{-r} + \left|\frac{1-z}{1 - Q(z)}\right|^r .$$

It therefore suffices to show that

$$(3.11) \qquad \left|\frac{1-z}{1 - Q(z)}\right| \le cn^{-1} \quad , \quad n \ge 2,$$

or

$$\left|\frac{n^2(1-z)^2}{n(1-z) - (1-z^n)}\right|^2 \le c^2 \quad , \quad n \ge 2,$$

where $c > 0$ is an absolute constant. Putting $z = e^{i\theta}$ we have

$$(3.12)\qquad |n^2(1-z)^2|^2 = 16n^4\sin^4(\theta/2) \le \theta^4 n^4.$$

Next, for $0 < \theta < 2\epsilon_r/n$, we have

$$|n(1-z) - (1-z^n)|^2 = 2n(n-1) + 2 - 2n(n-1)\cos\theta + 2n\cos n\theta$$
$$- 2\cos n\theta - 2n\cos(n-1)\theta$$

$$= 4n(n-1)\sin^2(\theta/2) + 4\sin^2(n\theta/2) - 4n\sin(\theta/2)\sin[(n-1/2)\theta]$$

$$= 4n^2\sin^2(\theta/2) + 4\sin^2(n\theta/2) - 8n\sin(\theta/2)\sin(n\theta/2)\cos[(n-1)\theta/2]$$

$$= 4\{[n\sin(\theta/2) - \sin(n\theta/2)]^2 + 4n\sin(\theta/2)\sin(n\theta/2)\sin^2[(n-1)\theta/4]\}$$

$$\ge 16n\sin(\theta/2)\sin(n\theta/2)\sin^2[(n-1)\theta/4]$$

$$\ge 16n(2/\pi)^4(\theta/2)(n\theta/2)[(n-1)\theta/4]^2.$$

Hence

$$(3.13)\qquad |n(1-z) - (1-z^n)|^2 \ge (4/\pi^4)\theta^4 n^2(n-1)^2.$$

The inequalities (3.12), (3.13) yield (3.11) with $c = \pi^2$. Hence,

$$|(1-z)^r - 1/p(z)| \le (1+\pi^{2r})n^{-r}, \quad 0 < \theta < 2\epsilon_r/n.$$

On choosing $\epsilon = \epsilon_r$ in Case 2 we conclude that

$$\max_{|z|\le 1} |(1-z)^r - 1/p(z)| \le c_r n^{-r}, \quad n = 1,2,3,\ldots,$$

where $c_r > 0$ depends only on r. Using a standard technique, the last inequality implies (3.1) for some constant B_r depending only on r (recall that $p(z)$ is of degree $(n-1)r^2 - r$).

To prove the upper bound in (1.4) we write

$$f(z) = g(z)\prod_{j=1}^{\nu}(z-z_j)^{r_j},$$

where $z_1, z_2, \ldots, z_\nu$ are the distinct zeros of f on $|z| = 1$ and g

is analytic in the closed disk $D: |z| \le 1$ and different from zero there. We just proved that

$$E_{on}((z-z_j)^{r_j};D) \le B_j n^{-r_j}.$$

Also, by Theorem 3, there exist constants $A > 0$ and $0 < \rho < 1$ such that

$$E_{on}(g;D) \le A\rho^n \quad , \quad n = 1,2,\ldots .$$

Applying Lemma 4.2 in [1] we conclude that for some constants $A_o > 0$ and $0 < \rho_o < 1$,

$$\begin{aligned} E_{on}(f;D) &\le \mathrm{const}(f) \sum_{j=1}^{\nu} n^{-r_j} + A_o\rho_o^n \\ &\le \mathrm{const}(f)\cdot n^{-r} \quad , \quad n = 1,2,3,\ldots , \end{aligned}$$

where $r = \min_j r_j$.

Next we prove the lower bound in (1.5). Pick a polynomial $P_n(z)$ of degree $\le n$ such that

$$\|(1-z)^r - 1/P_n(z)\|_D = E_{on}((1-z)^r;D) =: E_n \tag{3.14}$$

and let $p_n(\theta)$ denote the trigonometric polynomial $P_n(e^{i\theta})$. Then

$$\|(1-e^{i\theta})^r - 1/p_n(\theta)\|_{[-\pi,\pi]} = E_n \tag{3.15}$$

and therefore

$$|p_n(0)| \ge 1/E_n . \tag{3.16}$$

For $|\theta| \ge (\pi/2)(3E_n)^{1/r} =: \delta$ we have

$$|1-e^{i\theta}|^r = |2\sin(\theta/2)|^r \ge |2\theta/\pi|^r \ge 3E_n .$$

Hence (by (3.15))

(3.17) $$|p_n(\theta)| \le 1/(2E_n) \quad \text{for} \quad |\theta| \ge \delta.$$

It follows (see (3.16)) that $|p_n(\theta)|$ attains its maximum at some point θ_0 in $[-\delta,\delta]$. Now,

$$|p_n(\theta_0) - p_n(\delta)| \ge |p_n(\theta_0)| - |p_n(\delta)| = \|p_n\| - |p_n(\delta)| \ge \|p_n\| - 1/(2E_n),$$

where $\|\cdot\|$ denotes the sup norm on $[-\pi,\pi]$. Since $\|p_n\| \ge 1/E_n$ by (3.16), we obtain

(3.18) $$|p_n(\theta_0) - p_n(\delta)| \ge \|p_n\|/2.$$

On the other hand,

$$|p_n(\theta_0) - p_n(\delta)| \le |\theta_0 - \delta|\cdot\|p_n'\| \le 2\delta\|p_n'\| \le 2\delta n\|p_n\|$$

by Bernstein's inequality (see Lorentz [3, p.39]). Combining this with (3.18) we obtain that $\delta \ge 1/(4n)$. From the definition of δ it now follows that

$$E_n \ge c^r n^{-r},$$

where $0 < c < 1$ is an absolute constant. This proves the lower bound in (1.5).

For the general case, we pick a zero of f of the smallest order r ($z=1$, say) and write $f(z) = (1-z)^r(a+g(z))$, where $a \ne 0$ and $g(z) = O(1-z)$. We can find $\epsilon = \epsilon(f) > 0$ such that $|g(e^{i\theta})| < |a|/2$ for $|\theta| \le \epsilon$. Using the above argument (with obvious modifications) one can show that

$$\max_{-\epsilon\le\theta\le\epsilon} |(1-e^{i\theta})^r(a+g(e^{i\theta})) - 1/p_n(\theta)| \ge c(f)n^{-r}$$

which yields the lower bound in (1.4). □

References

1. Levin A.L. and Saff E.B., Degree of approximation of real functions by reciprocals of real and complex polynomials, SIAM J. Math. Analysis (to appear).

2. Levin A.L. and Saff E.B., Jackson type theorems in approximation by reciprocals of polynomials, Rocky Mountain Journal of Mathematics (to appear).

3. Lorentz G.G., Approximation of functions, Holt, Rinehart and Winston, New York, 1966.

4. Walsh J.L., On approximation to an analytic function by rational functions of best approximation, Math. Z., 38(1934), 163-176.

STRONG ASYMPTOTICS FOR L_p EXTREMAL POLYNOMIALS ($1 < p \leq \infty$) ASSOCIATED WITH WEIGHTS ON $[-1,1]$

by

D. S. Lubinsky[1]
National Research Institute for Mathematical Sciences
C.S.I.R.,
P.O. Box 395,
Pretoria 0001
Rep. of South Africa

E. B. Saff[2]
Institute for Constructive Mathematics
Department of Mathematics
University of South Florida
Tampa, FL 33620
USA

Abstract. While Szegö type asymptotics of orthonormal polynomials are classical, there has been a longstanding lack of corresponding results for L_p extremal polynomials, $p \neq 2$. In particular, in a 1969 paper, Widom raised the question of $p = \infty$. Here we fill some of the gaps for $1 < p \leq \infty$.

1. Introduction

Let $0 < p \leq \infty$, and $w \in L_p[-1,1]$ be non-negative in $[-1,1]$ and positive on a set of positive Lebesgue measure. We can then define for $n = 1,2,3,\ldots$,

---

[1]Part time at Dept. of Mathematics, Witwatersrand University, 1 Jan Smuts Avenue, Johannesburg 2001, South Africa.

[2]Research supported, in part, by the National Science Foundation Under Grant DMS-8620098.

AMS(MOS) Classification: Primary 41A60, 42C05.
Key Words and Phrases: Extremal polynomials, strong or power or Szegö asymptotics.

(1.1) $$E_{np}(w) := \inf_{P\in\mathcal{P}_{n-1}} \|\{x^n - P(x)\}w(x)\|_{L_p[-1,1]},$$

where $\mathcal{P}_{n-1}$ denotes the class of real polynomials of degree at most n-1. It is easily seen that there is at least one monic polynomial $T_{np}(w,x) = x^n + \cdots \in \mathcal{P}_n$ such that

(1.2) $$\|T_{np}(w,x)w(x)\|_{L_p[-1,1]} = E_{np}(w).$$

We call $T_{np}(w,x)$ an L_p <u>extremal polynomial</u> for w. We define also the <u>normalized extremal polynomials</u>

(1.3) $$p_{np}(w,x) := T_{np}(w,x)/E_{np}(w),$$

$n = 1,2,3,\ldots$, satisfying

(1.4) $$\|p_{np}(w,x)w(x)\|_{L_p[-1,1]} = 1.$$

When $p = 2$, $p_{np}(w,x)$ is just the orthonormal polynomial of degree n for the weight w^2.

This paper addresses the asymptotics of $T_{np}(w,x)$ in $\mathbb{C}\backslash[-1,1]$ as $n \to \infty$. Under general conditions on w, Fekete and Walsh [3] and Widom [11] established <u>nth root asymptotics</u>. For example, if $w(x) > 0$ a.e. in $[-1,1]$, their results imply that

(1.5) $$\lim_{n\to\infty} [T_{np}(w,z)]^{1/n} = \varphi(z)/2,$$

locally uniformly in $\mathbb{C}\backslash[-1,1]$, where

(1.6) $$\varphi(z) := z + \sqrt{z^2-1}, \quad z \in \mathbb{C}\backslash[-1,1],$$

is the usual conformal map of $\mathbb{C}\backslash[-1,1]$ onto $\{\zeta : |\zeta| > 1\}$. Here the branch of the nth root is chosen so that $[T_{np}(w,z)]^{1/n}$ behaves like z at ∞.

The asymptotics for $T_{np}(w,z)$ itself have proved more elusive. In his 1969 paper, Widom [12, p.205] remarked that even in the case of weights on $[-1,1]$. Szegö type asymptotics for $T_{n\infty}(w,z)$ had not yet been established. While Widom obtained asymptotics for $E_{n\infty}(w)$ and its analogue in more general situations than that treated here, he could not turn these into asymptotics for the polynomials. In this paper, we shall fill this gap, at least for $1 < p \leq \infty$, $p \neq 2$.

Of course for $p = 2$, everything is classical: Assuming the Szegö condition,

(1.7) $$\int_{-1}^{1} \log w(x)\, dx/\sqrt{1-x^2} > -\infty,$$

Szegö (see [10]) proved that locally uniformly in $\mathbb{C}\backslash[-1,1]$,

(1.8) $$\lim_{n\to\infty} p_{n2}(w,z)/\varphi(z)^n = (2\pi)^{-1/2} D^{-2}(F(\phi);\ \varphi(z)^{-1}),$$

where

(1.9) $$F(\phi) := w(\cos\phi)|\sin\phi|^{1/2}, \quad \phi \in \mathbb{R},$$

and $D(\cdot;\cdot)$ is the Szegö function

(1.10) $$D(F(\phi);\ u) := \exp\left[\frac{1}{4\pi}\int_{-\pi}^{\pi} \log F(\phi)\ \frac{1+ue^{-i\phi}}{1-ue^{-i\phi}}\, d\phi\right],$$

$|u| < 1$. Taking $z = \infty$ formally in (1.8), we see also that

(1.11) $$\lim_{n\to\infty} E_{n2}(w)2^n = (2\pi)^{1/2} D^2(F(\phi);\ 0)$$

$$= (2\pi)^{1/2} \exp\left[\frac{1}{2\pi}\int_{-\pi}^{\pi} \log F(\phi)d\phi\right]$$

$$= \pi^{1/2} G[w],$$

after some elementary manipulations, where

(1.12) $$G[w] := \exp\left[\pi^{-1}\int_{-1}^{1} \log w(x)dx/\sqrt{1-x^2}\right]$$

is a geometric mean of w. See [5,8] for recent reviews.

We have used the term "strong asymptotics" in our title to describe (1.8). Other commonly used names are power asymptotic, Szegö asymptotic, or full exterior asymptotic. We can now state our main result:

Theorem 1.1. Let $1 < p < \infty$ and $w \in L_p[-1,1]$ be a non-negative function such that for each $r < \infty$, $w^{-1} \in L_r[-1,1]$. Let

$$(1.13)\qquad \sigma_p := \{\Gamma(1/2)\Gamma((p+1)/2)/\Gamma(p/2+1)\}^{1/p},$$

and

$$(1.14)\qquad F_p(\phi) := w(\cos\phi)|\sin\phi|^{1/p}, \quad \phi \in \mathbb{R}.$$

Then

$$(1.15)\qquad \lim_{n\to\infty} E_{np}(w)2^{n-1+1/p} = \sigma_p G[w].$$

Furthermore, uniformly in closed subsets of $\mathbb{C}\setminus[-1,1]$, we have

$$(1.16)\qquad \lim_{n\to\infty} T_{np}(w,z)/\{\varphi(z)/2\}^n = D^{-2}(F_p(\phi);\ \varphi(z)^{-1})D^2(F_p(\phi);\ 0),$$

and

$$(1.17)\qquad \lim_{n\to\infty} p_{np}(w,z)/\varphi(z)^n = (2\sigma_p)^{-1}D^{-2}(F_p(\phi);\ \varphi(z)^{-1}).$$

For $p = \infty$, we shall prove:

Theorem 1.2. Let $w(x)$ be positive and continuous in $[-1,1]$, and let

$$(1.18)\qquad \sigma_\infty := 1,$$

and

$$(1.19)\qquad F_\infty(\phi) := w(\cos\phi), \quad \phi \in \mathbb{R}.$$

Then (1.15), (1.16) and (1.17) remain valid for $p = \infty$.

We note that our condition on w^{-1} in Theorem 1.1 implies Szegö's condition and severely restricts the zeros of w: It allows zeros of logarithmic, but not algebraic, strength. We shall prove a lim sup result corresponding to (1.15) under only Szegö's condition - see Theorem 2.2. If we could prove a matching lim inf result, then at least for $2 < p < \infty$, the Szegö-type asymptotics (1.16) and (1.17) would follow under only Szegö's condition (1.7).

This paper is organized as follows: In Section 2, we obtain asymptotics for $E_{np}(w)$ and in Section 3, we obtain the asymptotics for $T_{np}(w,z)$.

2. Asymptotics for $E_{np}(w)$.

First we list Bernstein's explicit formula for $E_{np}(w)$ and $T_{np}(w,x)$ for special w:

Lemma 2.1. *Let* q *be a positive integer and* $S(x)$ *be a polynomial of degree* $2q$, *positive in* $(-1,1)$, *possibly with simple zeros at* ± 1, *and let*

$$(2.1)\qquad V(x) := \{(1-x^2)/S(x)\}^{1/2}, \quad x \in (-1,1),$$

and for $0 < p \le \infty$,

$$(2.2)\qquad V_p(x) := (1-x^2)^{-1/(2p)} V(x), \quad x \in (-1,1).$$

Further, let σ_p *be defined by (1.13) for* $0 < p < \infty$ *and by (1.18) for* $p = \infty$, *and let the Szegö function and geometric mean be defined by (1.10) and (1.12) respectively.* *Let* $n \ge q$.

(a) *Then for* $1 \le p \le \infty$,

$$(2.3)\qquad E_{np}(V_p) = \sigma_p 2^{-n+1-1/p} G[V_p],$$

and for $0 < p < 1$,

$$(2.4)\qquad E_{np}(V_p) \le \sigma_p 2^{-n+1-1/p} G[V_p].$$

(b) *Let*

$$(2.5)\qquad \tau_n(x) := 2^{-n-1/p}G[V_p]\{z^{-n}D^{-2}(V(\cos\phi);z) + z^nD^{-2}(V(\cos\phi);z^{-1})\},$$

$x := \cos\theta$; $z := e^{i\theta}$; $\theta \in [0,\pi]$. _Then_ $\tau_n(x)$ _is a monic polynomial of degree_ n, _and for_ $0 < p \leq \infty$,

$$(2.6)\qquad \|\tau_n V_p\|_{L_p[-1,1]} = \sigma_p 2^{-n+1-1/p}G[V_p],$$

while for $1 \leq p \leq \infty$,

$$(2.7)\qquad T_{np}(V_p,x) = \tau_n(x).$$

Finally,

$$(2.8)\qquad |\tau_n(x)V(x)| \leq 2^{-n+1-1/p}G[V_p], \quad x \in [-1,1],$$

and for $u \in \mathbb{C}\backslash[-1,1]$,

$$(2.9)\qquad |\tau_n(u)/\{2^{-n-1/p}G[V_p]\varphi(u)^nD^{-2}(V(\cos\phi);\varphi(u)^{-1})\} - 1|$$

$$\leq |\varphi(u)|^{2q-2n-2}.$$

Proof. See Theorem 13.1 in [7], which is just a reformulation of statements in Achieser [1,pp.250-4]. □

We can now prove:

Theorem 2.2. _Let_ $0 < p < \infty$ _and_ $w \in L_p[-1,1]$ _be a non-negative function_. _Then_

$$(2.10)\qquad \limsup_{n\to\infty} E_{np}(w)2^{n-1+1/p} \leq \sigma_p G[w],$$

where, if the integral in the definition (1.12) of $G[w]$ _diverges to_ $-\infty$, _we interpret_ $G[w]$ _as_ 0.

Proof. We remark first that $G[w] < \infty$ is an easy consequence of the arithmetic-geometric mean inequality (cf.[10]) and the fact that $w \in L_p[-1,1]$ implies $w \in L_s[-1,1]$ for $s < p$. For a given n, and

$0 < p < \infty$, let $V_p(x)$ be a function given by (2.2), fulfilling the hypotheses of Lemma 2.1. Let $\epsilon \in (0,1)$ and $w_\epsilon(x) := \max\{w(x), \epsilon\}$, $x \in [-1,1]$. Further let $\mathcal{F}$ be a measurable subset of $[-1,1]$ with $\|w\|_{L_\infty(\mathcal{F})} < \infty$, and let $\mathcal{E} := [-1,1]\setminus\mathcal{F}$. With the notation of Lemma 2.1, we have, for $n \geq q$,

$$(2.11)\quad E^p_{np}(w) \leq E^p_{np}(w_\epsilon) \leq \|\tau_n w_\epsilon\|^p_{L_p[-1,1]} = \|\tau_n w_\epsilon\|^p_{L_p(\mathcal{F})} + \|\tau_n w_\epsilon\|^p_{L_p(\mathcal{E})}$$

$$\leq \|\tau_n V_p\|^p_{L_p(\mathcal{F})} \|V_p^{-1} w_\epsilon\|^p_{L_\infty(\mathcal{F})}$$

$$+ \{2^{-n+1-1/p} G[V_p]\}^p \|V^{-1} w_\epsilon\|^p_{L_p(\mathcal{E})}$$

$$\leq \{\sigma_p 2^{-n+1-1/p} G[V_p]\}^p$$

$$\times \left\{\|V_p^{-1} w_\epsilon\|^p_{L_\infty(\mathcal{F})} + \sigma_p^{-p} \|V^{-1} w_\epsilon\|^p_{L_p(\mathcal{E})}\right\}.$$

Now taking $S(x) := (1-x^2)R^2(x)$ in Lemma 2.1, where $R(x)$ is a polynomial positive in $[-1,1]$, we see that

$$(V_p^{-1} w_\epsilon)(x) = (1-x^2)^{1/(2p)} R(x) w_\epsilon(x)$$

and

$$(V^{-1} w_\epsilon)(x) = R(x) w_\epsilon(x).$$

Let $g(x)$ be a continuous positive function in $[-1,1]$. We can choose a sequence $R = R_{n-1} \in \mathcal{P}_{n-1}$ of polynomials converging uniformly in $[-1,1]$ to g as $n \to \infty$. Then (2.11) yields

$$(2.12)\quad \limsup_{n\to\infty} \left\{E_{np}(w) 2^{n-1+1/p}\right\}^p$$

$$\leq \sigma_p^p G[(1-x^2)^{-1/(2p)} g(x)^{-1}]^p$$

$$\times \left\{\|(1-x^2)^{1/(2p)} g(x) w_\epsilon(x)\|^p_{L_\infty(\mathcal{F})} + \sigma_p^{-p} \|g w_\epsilon\|^p_{L_p(\mathcal{E})}\right\}.$$

Next, we claim that (2.12) holds more generally for any measurable function $g(x)$ that is bounded above and below by positive constants. To see this, note first that for such a $g(x)$, we can choose continuous functions $g_m(x)$, $m = 1,2,\dots$, bounded above and below by positive constants independent of n, such that

$$\lim_{m\to\infty} g_m(x) = g(x) \qquad \text{a.e. in } [-1,1].$$

For example, we can choose

$$g_m(x) := \int_{[x-1/m,x+1/m]\cap[-1,1]} g(t)dt \Big/ \int_{[x-1/m,x+1/m]\cap[-1,1]} dt.$$

Furthermore, we can choose a measurable set $\hat{\mathcal{F}} \subset \mathcal{F}$ such that $\operatorname{meas}(\mathcal{F}\setminus\hat{\mathcal{F}})$ is as small as we please and

$$\lim_{m\to\infty} g_m(x) = g(x) \qquad \text{uniformly on } \hat{\mathcal{F}}.$$

Let $\hat{\mathcal{E}} := [-1,1]\setminus\hat{\mathcal{F}}$. As g_m and g are bounded above and below by positive constants independent of n, inequality (2.12) yields for each m,

$$\text{(2.13)} \qquad \limsup_{n\to\infty}\left\{E_{np}(w)2^{n-1+1/p}\right\}^p$$
$$\le \sigma_p^p\, G[(1-x^2)^{-1/(2p)}g_m(x)^{-1}]^p$$
$$\times \left\{\|(1-x^2)^{1/(2p)}g_m(x)w_\epsilon(x)\|^p_{L_\infty(\hat{\mathcal{F}})} + \sigma_p^{-p}\|g_m w_\epsilon\|^p_{L_p(\hat{\mathcal{E}})}\right\}.$$

and so, on letting $m \to \infty$, we get

$$\limsup_{n\to\infty}\left\{E_{np}(w)2^{n-1+1/p}\right\}^p$$
$$\le \sigma_p^p\, G[(1-x^2)^{-1/(2p)}g(x)^{-1}]^p$$
$$\times \left\{\|(1-x^2)^{1/(2p)}g(x)w_\epsilon(x)\|^p_{L_\infty(\hat{\mathcal{F}})} + \sigma_p^{-p}\|g w_\epsilon\|^p_{L_p(\hat{\mathcal{E}})}\right\}.$$

Since we can choose $\hat{\mathcal{F}} \subset \mathcal{F}$ such that $\text{meas}(\hat{\mathcal{E}}\backslash\mathcal{E})$ is as small as desired, we obtain via Lebesgue's Dominated Convergence Theorem, that (2.12) holds (as claimed) for any measurable g that is bounded above and below by positive constants.

Now, we can choose $\mathcal{F}$ to omit small intervals containing -1 and 1. Then $(1-x^2)^{1/(2p)}w_\epsilon(x)$ is bounded above and below by positive constants on $\mathcal{F}$, and the same is true for

$$g(x) := \begin{cases} (1-x^2)^{-1/(2p)} w_\epsilon(x)^{-1}, & x \in \mathcal{F} \\ 1, & x \in \mathcal{E}. \end{cases}$$

Thus, if $\chi_{\mathcal{F}}$ and $\chi_{\mathcal{E}}$ denote the characteristic functions of $\mathcal{F}$ and $\mathcal{E}$ respectively, we obtain from (2.12)

$$\text{(2.14)} \qquad \limsup_{n\to\infty} \left\{E_{np}(w)2^{n-1+1/p}\right\}^p$$

$$\le \sigma_p^p \, G[w_\epsilon(x)\chi_{\mathcal{F}}(x) + (1-x^2)^{-1/(2p)}\chi_{\mathcal{E}}(x)]^p$$

$$\times \left\{1 + \sigma_p^{-p}\|w_\epsilon\|^p_{L_p(\mathcal{E})}\right\}.$$

Choosing $\mathcal{F}$ such that $\text{meas}(\mathcal{E})$ is as small as desired, (2.14) yields

$$\limsup_{n\to\infty} \left\{E_{np}(w)2^{n-1+1/p}\right\}^p \le \sigma_p^p \, G[w_\epsilon]^p.$$

Finally, (2.10) follows by writing

$$\log w_\epsilon^{-1} = \log^+ w_\epsilon^{-1} - \log^+ w_\epsilon ,$$

and observing that since $\log^+ w_\epsilon(\cos\phi)$ is bounded above by the integrable function $\log^+(w(\cos\phi) + 1)$ for $\epsilon < 1$, we have

$$\lim_{\epsilon\to 0+} \int_0^\pi \log^+ w_\epsilon(\cos\phi)d\phi = \int_0^\pi \log^+ w(\cos\phi)d\phi$$

and, by the Monotone Convergence Theorem,

$$\lim_{\epsilon\to 0+} \int_0^\pi \log^+ w_\epsilon(\cos\phi)^{-1} d\phi = \int_0^\pi \log^+ w(\cos\phi)^{-1} d\phi. \qquad \square$$

<u>Theorem 2.3</u>. <u>Let</u> $w(x)$ <u>be a bounded non-negative Riemann integrable function on</u> $[-1,1]$. <u>Then</u>

$$\limsup_{n\to\infty} E_{n\infty}(w)2^{n-1} \leq G[w]. \tag{2.15}$$

<u>Proof</u>. Let V_∞ be a function given by (2.2), fulfilling the hypotheses of Lemma 2.1. We obtain

$$E_{n\infty}(w) \leq \|\tau_n V_\infty\|_{L_\infty[-1,1]} \|V_\infty^{-1} w\|_{L_\infty[-1,1]} = 2^{-n+1} G[V_\infty] \|V_\infty^{-1} w\|_{L_\infty[-1,1]} \cdot$$

by (2.6) and (1.18). Choosing $S(x) := (1-x^2)R^2(x)$ as before, and then choosing $R(x)$ to approximate the reciprocal of a continuous positive function g, we obtain

$$\limsup_{n\to\infty} E_{n\infty}(w)2^{n-1} \leq G[g] \|g^{-1} w\|_{L_\infty[-1,1]} \cdot$$

Since w is Riemann integrable, a theorem of M. Riesz on one-sided approximation [4,p.73] ensures that we can choose a continuous function $g = g_\delta$ (even a polynomial) such that

$$w(x) \leq g(x) \quad \text{in} \quad [-1,1]$$

and

$$\int_{-1}^1 (g(x)-w(x))dx < \delta,$$

for any given $\delta > 0$. This g yields the desired result if w has a positive lower bound in $[-1,1]$. When the latter condition fails, replace w by w_ϵ as in the previous proof, and then let $\epsilon \to 0+$. $\square$

We now turn to the corresponding asymptotic lower bounds. Together with Theorems 2.2 and 2.3, they immediately yield (1.15).

Lemma 2.4. Let $1 < p < \infty$ and $w \in L_p[-1,1]$ be a non-negative function such that $w^{-1} \in L_r[-1,1]$ for every $r < \infty$. Then

$$\text{(2.16)} \qquad \liminf_{n\to\infty} E_{np}(w) 2^{n-1+1/p} \geq \sigma_p G[w].$$

If $p = 1$ or $p = \infty$, (2.16) remains valid provided w is positive and continuous in $[-1,1]$.

Proof. Let $1 < p < \infty$ and $r > s > 1$, with $r^{-1} + s^{-1} = 1$, and $s \leq p$. An easy consequence of Hölder's inequality is that if $H^{-1} \in L_{pr/s}[-1,1]$ and $JH \in L_p[-1,1]$, then

$$\text{(2.17)} \qquad \|JH\|_{L_p[-1,1]} \geq \|J\|_{L_{p/s}[-1,1]} \|H^{-1}\|^{-1}_{L_{pr/s}[-1,1]}$$

(see [6, Lemma 3.1]). Let $V_{p/s}(x)$ be a function given by (2.2), fulfilling the hypotheses of Lemma 2.1. Applying (2.17) with $J(x) := \{x^n - P(x)\} V_{p/s}(x)$ and $H(x) := V_{p/s}(x)^{-1} w(x)$, we obtain

$$E_{np}(w) \geq E_{n,p/s}(V_{p/s}) \left\| V_{p/s} w^{-1} \right\|^{-1}_{L_{pr/s}[-1,1]}$$

$$= 2^{-n+1-s/p} \sigma_{p/s} G[V_{p/s}] \left\| V_{p/s} w^{-1} \right\|^{-1}_{L_{pr/s}[-1,1]} ,$$

by (2.3) and as $p/s \geq 1$. Choosing $S(x) := R^2(x)$, where $R(x)$ does not vanish in $[-1,1]$, and choosing $R(x)$ to uniformly approximate a function $g(x)$ positive and continuous in $[-1,1]$, we obtain

$$\liminf_{n\to\infty} E_{np}(w) 2^{n-1+s/p} \geq$$

$$\sigma_{p/s} G[(1-x^2)^{(1-s/p)/2} g(x)^{-1}] \|(1-x^2)^{(1-s/p)/2} g(x)^{-1} w(x)^{-1}\|^{-1}_{L_{pr/s}[-1,1]} .$$

Choosing $g(x)$ to approximate $(1-x^2)^{(1-s/p)/2} w(x)^{-1}$ in a suitable sense, we obtain

$$\liminf_{n\to\infty} E_{np}(w) 2^{n-1+s/p} \geq \sigma_{p/s} G[w] \|1\|^{-1}_{L_{pr/s}[-1,1]} .$$

Here $\|1\|_{L_{pr/s}[-1,1]} = 2^{s/(pr)} \to 1$ as $r \to \infty$, and also then $s \to 1$, so $\sigma_{p/s} \to \sigma_p$. Then (2.16) follows.

Finally in the cases $p = 1, \infty$, the proof is much easier: Since w is positive and continuous, one can choose a sequence $\{R_n\}_1^\infty$ of polynomials such that

$$R_n w \to 1 \quad \text{as} \quad n \to \infty,$$

uniformly in $[-1,1]$. By suitable choice of V_1 and V_∞, we easily obtain (2.16). □

We remarked in Section 1 that our conditions on w^{-1} imply Szegö's condition (1.7), and we now briefly justify this. Note that then also $G[w] > 0$. Suppose that for some $s > 0$, $w^{-1} \in L_s[-1,1]$. By the arithmetic-geometric mean inequality [10,p.2],

$$\exp\left[\pi^{-1} \int_{-1}^{1} \log w^{-s/3}(x)dx/\sqrt{1-x^2}\right] \le \pi^{-1} \int_{-1}^{1} w^{-s/3}(x)dx/\sqrt{1-x^2}$$

$$\le \pi^{-1} \|w^{-1}\|^{s/3}_{L_s[-1,1]} \|(1-x^2)^{-1/2}\|_{L_{3/2}[-1,1]} < \infty.$$

Hence (1.7).

3. Asymptotics for Extremal Polynomials.

The main new ideas of this paper are contained in Lemmas 3.1 and 3.2, where standard L_2 techniques [8] are turned into L_p ones:

Lemma 3.1. *Let* $2 \le p \le \infty$ *and* $w \in L_p[-1,1]$ *be a non-negative function that is positive on a set of positive measure. Let* $n \ge 1$, $P(z)$ *be a polynomial of degree* n *with leading coefficient* Δ, *and let*

$$(3.1) \qquad A := E_{np}(w)\Delta.$$

Then for $z \in \mathbb{C}\setminus[-1,1]$,

(3.2) $$|P(z)/p_{np}(w,z)-1| \le d(z)^{-1}\left\{\|Pw\|^2_{L_p[-1,1]} - A^2\right\}^{1/2} + |A-1|,$$

<u>where</u> $d(z)$ <u>denotes the distance from</u> z <u>to</u> $[-1,1]$.

<u>Proof</u>. Suppose first $p < \infty$. Note that as $p_{np}(w,x)$ is an extremal polynomial for w^p, we have

$$\int_{-1}^{1} |p_{np}(w,x)|^{p-2} p_{np}(w,x)\pi(x)w^p(x)dx = 0,$$

for each $\pi \in \mathcal{P}_{n-1}$ (cf.[2,p.9]). Let

$$\hat{w}(x) := |p_{np}(w,x)|^{p-2} w^p(x), \quad x \in [-1,1].$$

By its normalization, we see that $p_{np}(w,x)$ is the orthonormal polynomial of degree n for $\hat{w}(x)$, that is

$$\int_{-1}^{1} p_{np}(w,x)^2 \hat{w}(x)dx = 1,$$

and

$$\int_{-1}^{1} p_{np}(w,x)\pi(x)\hat{w}(x)dx = 0, \quad \pi \in \mathcal{P}_{n-1}.$$

Let

$$q(x) := P(x) - Ap_{np}(w,x),$$

a polynomial of degree $\le n-1$. Let $\ell_{jn}(x)$, $j = 1,2,\dots,n$, denote the fundamental polynomials of Lagrange interpolation at the zeros $x_{1n}, x_{2n}, \dots, x_{nn}$ of $p_{np}(w,x)$. We have

$$q(x) = \sum_{j=1}^{n} q(x_{jn})\ell_{jn}(x).$$

By a well-known formula [8,p.6], [4,p.114, eqn.(6.3)], we have

$$\ell_{jn}(x) = \lambda_{jn} p_{n-1}(x_{jn})(\gamma_{n-1}/\gamma_n) p_{np}(w,x)/(x-x_{jn}),$$

where λ_{jn}, $j = 1,2,\ldots,n$, are the Gauss-Christoffel numbers of order n for $\hat{w}$, while $p_{n-1}(x)$ is the orthonormal polynomial of degree $n-1$ for $\hat{w}$ and γ_{n-1} and γ_n are the leading coefficients of $p_{n-1}(x)$ and $p_{np}(w,x)$, respectively. Here, as $\hat{w}$ is a weight on $[-1,1]$,

$$\gamma_{n-1}/\gamma_n \leq 1$$

[4,p.41] and so we obtain for $z \in \mathbb{C}\backslash[-1,1]$,

$$(3.3) \qquad |P(z)/p_{np}(w,z) - A| = |q(z)/p_{np}(w,z)|$$

$$\leq \sum_{j=1}^{n} \lambda_{jn} |p_{n-1}(x_{jn})| |q(x_{jn})| / |z-x_{jn}|$$

$$\leq d(z)^{-1} \left\{ \sum_{j=1}^{n} \lambda_{jn} p_{n-1}^2(x_{jn}) \right\}^{1/2} \left\{ \sum_{j=1}^{n} \lambda_{jn} q^2(x_{jn}) \right\}^{1/2}$$

$$= d(z)^{-1} \left\{ \int_{-1}^{1} q^2(x)\hat{w}(x)dx \right\}^{1/2},$$

by orthonormality, and the exactness of the Gauss quadrature formula. Here

$$I := \int_{-1}^{1} q^2(x)\hat{w}(x)dx = \int_{-1}^{1} P^2(x)\hat{w}(x)dx - 2A\int_{-1}^{1} P(x)p_{np}(w,x)\hat{w}(x)dx + A^2$$

$$= \int_{-1}^{1} P^2(x)\hat{w}(x)dx - A^2,$$

since A is the coefficient of $p_{np}(w,x)$ in the expansion of $P(x)$ in the orthonormal polynomials for $\hat{w}$. Taking account of the definition of $\hat{w}$ and using Hölder's inequality with parameters $2/p$ and $1-2/p$, we obtain

$$I \leq \|Pw\|^2_{L_p[-1,1]} \|p_{np}(w,x)w(x)\|^{p-2}_{L_p[-1,1]} - A^2 = \|Pw\|^2_{L_p[-1,1]} - A^2.$$

Then (3.2) follows from (3.3).

Finally, if $p = \infty$, we can simply let $p \to \infty$ in (3.2), noting that (as is well-known) $E_{np}(w)$ and $p_{np}(w,x)$ respectively converge to $E_{n\infty}(w)$ and $p_{n\infty}(w,x)$ as $p \to \infty$. □

<u>Proof of (1.16) and (1.17) when</u> $2 \le p \le \infty$. Suppose first $p < \infty$. Let V_p be a function given by (2.2), fulfilling the hypotheses of Lemma 2.1. Note that by (2.3), (2.7) and (2.8),

$$(3.4) \qquad |p_{np}(V_p,x)V(x)| \le \sigma_p^{-1}, \qquad x \in [-1,1],$$

while by (2.3), (2.7) and (2.9), for $u \in \mathbb{C}\setminus[-1,1]$,

$$(3.5) \qquad |p_{np}(V_p,u)/\{(2\sigma_p)^{-1}\varphi(u)^n D^{-2}(V(\cos\phi);\ \varphi(u)^{-1})\} - 1|$$

$$\le |\varphi(u)|^{2q-2n-2},$$

where $2q$ is the degree of S. Let $\mathcal{F} \subset [-1,1]$ be a measurable set for which

$$\|w\|_{L_\infty(\mathcal{F})} < \infty,$$

and let $\mathcal{E} := [-1,1]\setminus\mathcal{F}$. Further, let

$$A := E_{np}(w)/E_{np}(V_p).$$

Substituting $P := p_{np}(V_p)$ in (3.2) yields for $z \in \mathbb{C}\setminus[-1,1]$,

$$(3.6) \qquad |p_{np}(V_p,z)/p_{np}(w,z) - 1|$$

$$\le d(z)^{-1}\left\{\|p_{np}(V_p,x)w(x)\|^2_{L_p[-1,1]} - A^2\right\}^{1/2} + |A-1|$$

$$\le d(z)^{-1}\left\{\left[\|p_{np}(V_p,x)V_p(x)\|_{L_p(\mathcal{F})}\|V_p^{-1}w\|_{L_\infty(\mathcal{F})}\right.\right.$$

$$\left.\left. + \sigma_p^{-1}\|V^{-1}w\|_{L_p(\mathcal{E})}\right]^2 - A^2\right\}^{1/2} + |A-1|$$

$$\le d(z)^{-1}\left\{\left[\|V_p^{-1}w\|_{L_\infty(\mathcal{F})} + \sigma_p^{-1}\|V^{-1}w\|_{L_p(\mathcal{E})}\right]^2 - A^2\right\}^{1/2} + |A-1|.$$

A glance at the right-hand side of (2.11) indicates its similarity to this last right-hand side and we proceed in a like fashion. Take $S(x) := (1-x^2)R^2(x)$ and $R = R_n(x)$ where $R_n(x)$ has degree $q-1 = q_n-1 \le (n-2)/2$ and $R_n(x) \to g(x)$ uniformly in $[-1,1]$ as $n \to \infty$, where $g(x)$ is positive and continuous in $[-1,1]$. Set

$$h_n(f,z) := (2\sigma_p)^{-1}\varphi(z)^n D^{-2}(f(\cos\phi);\ \varphi(z)^{-1}),$$

and with $f_p(x) := w(x)(1-x^2)^{1/(2p)}$ (so that $f_p(\cos\phi) = F_p(\phi)$), write

$$\frac{h_n(f_p,z)}{p_{np}(w,z)} = \alpha_n\beta_n\lambda_n, \tag{3.7}$$

where

$$\alpha_n = \alpha_n(V_p,z) := \frac{p_{np}(V_p,z)}{p_{np}(w,z)}, \quad \beta_n = \beta_n(V_p,z) := \frac{h_n(V,z)}{p_{np}(V_p,z)},$$

$$\lambda_n = \lambda_n(V_p,z) := \frac{h_n(f_p,z)}{h_n(V,z)}.$$

Then, for $V(x) = R_n(x)^{-1}$, $V_p(x) = R_n(x)^{-1}(1-x^2)^{-1/(2p)}$, inequality (3.5) together with the fact that $q \le n/2$ imply

$$\lim_{n\to\infty} \beta_n = 1, \tag{3.8}$$

uniformly in closed subsets of $\mathbb{C}\setminus[-1,1]$. Also, since $V \to g^{-1}$ uniformly on $[-1,1]$, we see from (1.10) that

$$\lim_{n\to\infty} \lambda_n = D^{-2}((f_p g)(\phi);\ \varphi(z)^{-1}), \tag{3.9}$$

uniformly in closed subsets of $\mathbb{C}\setminus[-1,1]$. Furthermore, from (1.15) and (3.6) we get for any closed set $K \subset \mathbb{C}\setminus[-1,1]$,

(3.10) $$\limsup_{n\to\infty} \|\alpha_n - 1\|_{L_\infty(K)} \le \delta_K\, B(g,\mathcal{F},K),$$

where $\delta_K := \|d(z)^{-1}\|_{L_\infty(K)}$ and

$$B(g,\mathcal{F},K) := \left\{\left[\|gf_p\|_{L_\infty(\mathcal{F})} + \sigma_p^{-1}\|gw\|_{L_p(\mathcal{E})}\right]^2 - G[gf_p]^2\right\}^{1/2}$$

$$+ |G[gf_p] - 1|.$$

Since

$$|\alpha_n\beta_n\lambda_n - 1| \le |\alpha_n - 1||\beta_n - 1||\lambda_n - 1| + |\alpha_n||\beta_n - 1| + |\beta_n||\lambda_n - 1| + |\lambda_n||\alpha_n - 1|,$$

it follows from (3.7) - (3.10) that

(3.11) $$\limsup_{n\to\infty} \left\|\frac{h_n(f_p,z)}{p_{np}(w,z)} - 1\right\|_{L_\infty(K)}$$

$$\le \|D^{-2}((f_p g)(\phi);\ \varphi(z)^{-1}) - 1\|_{L_\infty(K)}$$

$$+ \|D^{-2}((f_p g)(\phi);\ \varphi(z)^{-1})\|_{L_\infty(K)}\,\delta_K B(g,\mathcal{F},K).$$

Much as in the proof of Theorem 2.2, we note that (3.11) holds, more generally, when g is any measurable function bounded above and below by positive constants. Choosing $\mathcal{F}$ such that f_p is bounded above and below by positive constants in $\mathcal{F}$, we take

$$g(x) := \begin{cases} f_p(x)^{-1}, & x \in \mathcal{F} \\ 1 & ,\quad x \in \mathcal{E}. \end{cases}$$

Since we can choose $\text{meas}(\mathcal{E})$ as small as desired (note that $w(x)$ is positive a.e. in $[-1,1]$), we get $B(g,\mathcal{F},K) \to 0$ and $D^{-2}((f_p g)(\phi);\ \varphi(z)^{-1}) \to 1$ uniformly on K, as $\text{meas}(\mathcal{E}) \to 0$. Then (3.11) yields (1.17).

Finally, if $p = \infty$, the proof is substantially easier, since then our hypotheses on w ensure that we can choose $V_\infty = V_{\infty,n}$ in

Lemma 2.1 such that $V_\infty \to w$ uniformly in $[-1,1]$ as $n \to \infty$. □

Next, we turn to the cases $1 < p < 2$. The method below works also for $2 \leq p \leq \infty$ under the hypotheses of Theorems 1.1 and 1.2, but we preferred to include Lemma 3.1 because of its greater potential: it requires little more than asymptotics for $E_{np}(w)$, whereas Lemma 3.2 places awkward restrictions on w^{-1}:

Lemma 3.2. *Let* $1 < p < 2 < q$ *satisfy* $p^{-1} + q^{-1} = 1$ *and let* u, w *be non-negative functions on* $[-1,1]$ *that are positive on a set of positive measure and satisfy* $u \in L_1[-1,1]$ *and* $uw^{-1} \in L_q[-1,1]$. *Let* $\{p_n\}_0^\infty$ *denote the orthonormal polynomials for* u *satisfying*

$$\int_{-1}^{1} p_n(x)p_m(x)u(x)dx = \delta_{mn}, \quad m,n = 0,1,2,\ldots \; ,$$

Let $P(z)$ *be a polynomial of degree* n, *with leading coefficient* Λ, *and let* A *and* $d(z)$ *be as in Lemma 3.1*. *Let*

$$\Psi(x) := \left[((x^p+1)/2)^{1/(p-1)} - 1 \right]^{(p-1)/p}, \quad x \in [1,\infty). \tag{3.12}$$

Then for $z \in \mathbb{C}\backslash[-1,1]$,

$$\begin{aligned} &|P(z)/p_{np}(w,z) - 1| \\ &\leq |A-1| + 2d(z)^{-1}|p_n(z)/p_{np}(w,z)|\,|A|\Psi\left[\|Pw\|_{L_p[-1,1]}A^{-1}\right] \\ &\quad \times \{\|p_{n-1}uw^{-1}\|_{L_q[-1,1]} + d(z)^{-1}\|p_n uw^{-1}\|_{L_q[-1,1]}\}. \end{aligned} \tag{3.13}$$

Proof. We shall use Clarkson's inequalities [2,p.3] for $1 < p < 2$: For $f,g \in L_p[-1,1]$,

$$\begin{aligned} &\|f+g\|_{L_p[-1,1]}^{p/(p-1)} + \|f-g\|_{L_p[-1,1]}^{p/(p-1)} \\ &\leq 2\left\{\|f\|_{L_p[-1,1]}^{p} + \|g\|_{L_p[-1,1]}^{p}\right\}^{1/(p-1)}. \end{aligned}$$

Letting $\hat{P} := P/\Delta$, $f := \hat{P}w$ and $g := T_{np}(w,\cdot)w$, we note that

$$\|f+g\|_{L_p[-1,1]} = 2\|(f+g)/2\|_{L_p[-1,1]} \geq 2E_{np}(w),$$

and so Clarkson's inequalities yield

$$\left\|(\hat{P} - T_{np}(w))w\right\|_{L_p[-1,1]}^{p/(p-1)}$$

$$\leq 2\left\{\|\hat{P}w\|_{L_p[-1,1]}^p + E_{np}^p(w)\right\}^{1/(p-1)} - (2E_{np}(w))^{p/(p-1)},$$

and so using (3.1) and (3.12), we obtain

$$(3.14) \qquad \|(\hat{P} - T_{np}(w))w\|_{L_p[-1,1]} \leq 2E_{np}(w)\Psi\left[\|Pw\|_{L_p[-1,1]}A^{-1}\right].$$

Next, let γ_j denote the leading coefficient of $p_j(x)$, and

$$K_n(x,t) := \{\gamma_{n-1}/\gamma_n\}\{p_n(x)p_{n-1}(t) - p_n(t)p_{n-1}(x)\}/(x-t).$$

As in Lemma 3.1, $\gamma_{n-1}/\gamma_n \leq 1$, and so for $t \in [-1,1]$ and $x \in \mathbb{C}\backslash[-1,1]$,

$$(3.15) \qquad |K_n(x,t)| \leq d(x)^{-1}\{|p_n(x)p_{n-1}(t)| + |p_n(t)p_{n-1}(x)|\}.$$

Further as $\hat{P}(x) - T_{np}(w,x)$ has degree $\leq n-1$, we have [4,Ch.1],

$$\hat{P}(x) - T_{np}(w,x) = \int_{-1}^{1}(\hat{P}(t) - T_{np}(w,t))K_n(x,t)u(t)dt.$$

Then (3.15) and Hölder's inequality yield for $x \in \mathbb{C}\backslash[-1,1]$,

$$(3.16) \qquad |\hat{P}(x) - T_{np}(w,x)| \leq \|(\hat{P} - T_{np}(w))w\|_{L_p[-1,1]}$$

$$\times\, d(x)^{-1}\left\{|p_n(x)|\|p_{n-1}uw^{-1}\|_{L_q[-1,1]} + |p_{n-1}(x)|\|p_nuw^{-1}\|_{L_q[-1,1]}\right\}$$

Here, much as in the proof of Lemma 3.1,

$$p_{n-1}(x) = \frac{\gamma_{n-1}}{\gamma_n} \sum_{j=1}^{n} \lambda_{jn} p_{n-1}^2(x_{jn}) p_n(x)/(x-x_{jn}),$$

where $\{x_{jn}\}_1^n$ are the zeros of p_n and $\{\lambda_{jn}\}_1^n$ are the Gauss quadrature weights of order n for u. Then for $x \in \mathbb{C}\setminus[-1,1]$,

$$|p_{n-1}(x)/p_n(x)| \le d(x)^{-1} \sum_{j=1}^{n} \lambda_{jn} p_{n-1}^2(x_{jn}) = d(x)^{-1}.$$

Together with (3.14) and (3.16), this yields for $x \in \mathbb{C}\setminus[-1,1]$,

$$(3.17) \qquad |\hat{P}(x) - T_{np}(w,x)| \le 2d(x)^{-1} E_{np}(w) \Psi\left\{\|Pw\|_{L_p[-1,1]} A^{-1}\right\}$$

$$\times |p_n(x)| \left\{\|p_{n-1}uw^{-1}\|_{L_q[-1,1]} + d(x)^{-1} \|p_n uw^{-1}\|_{L_q[-1,1]}\right\}.$$

Finally,

$$|P(x)/p_{np}(w,x) - 1| = |A\hat{P}(x) - T_{np}(w,x)|/|T_{np}(w,x)|$$

$$\le |A| |\hat{P}(x) - T_{np}(w,x)|/|T_{np}(w,x)| + |A - 1|.$$

Substituting the estimate (3.17) into this last inequality yields (3.13). □

The difficulty above is choosing u so that $\|p_k uw^{-1}\|_{L_q[-1,1]}$ is bounded independent of k. Note that if

$$(3.18) \qquad u(x) := (1-x^2)^{1/2}, \quad x \in [-1,1],$$

then [4, p.35],

$$(3.19) \qquad \|p_n u\|_{L_\infty[-1,1]} \le (2/\pi)^{1/2}, \quad n = 1,2,3,\ldots .$$

<u>Proof of (1.16) and (1.17) when</u> $1 < p < 2$. Choose u by (3.18), and let V_p be a function given by (2.2), fulfilling the hypotheses of Lemma 2.1. Substituting $P(x) := p_{np}(V_p,x)$ in (3.13) and using

(3.19), we obtain

$$|p_{np}(V_p,z)/p_{np}(w,z) - 1|$$

$$\le |A-1| + 2d(z)^{-1}|p_n(z)/p_{np}(w,z)||A|\Psi\left[\|p_{np}(V_p)w\|_{L_p[-1,1]}A^{-1}\right]$$

$$\times (2/\pi)^{1/2}\|w^{-1}\|_{L_q[-1,1]}\{1 + d(z)^{-1}\}.$$

where $A = E_{np}(w)/E_{np}(V_p)$. Much as in the previous proof, we can choose a sequence of functions V_p such that $A \to 1$, $n \to \infty$, and

$$\|p_{np}(V_p)w\|_{L_p[-1,1]} \to 1, \quad n \to \infty.$$

Since $\Psi(1) = 0$, the Szegö type asymptotics for $p_n(z)$ and the above estimates easily yield (1.17) and hence (1.16). □

References

1. N.I. Achiezer, Theory of Approximation, (transl. by C.J. Hyman), Ungar, New York, 1956.

2. D. Braess, Nonlinear Approximation Theory, Springer Series in Computational Mathematics, Volume 7, Springer, Berlin, 1986.

3. M. Fekete and J.L. Walsh, On the Asymptotic Behaviour of Polynomials with Extremal Properties, and of Their Zeros, J. Analyse Math., 4(1954/5), 49-87.

4. G. Freud, Orthogonal Polynomials, Akademiai Kiado/Pergamon Press, Budapest, 1971.

5. D.S. Lubinsky, A Survey of General Orthogonal Polynomials for Weights on Finite and Infinite Intervals, to appear in Acta Appl. Math.

6. D.S. Lubinsky and E.B. Saff, Sufficient Conditions for Asymptotics Associated with Weighted Extremal Problems on ℝ, to appear in Rocky Mountain J. Math.

7. D.S. Lubinsky and E.B. Saff, Strong Asymptotics for Extremal Errors and Extremal Polynomials Associated with Weights on (-∞,∞), to appear in Springer Lecture Notes in Mathematics.

8. P. Nevai, Orthogonal Polynomials, Memoirs Amer. Math. Soc., Vol. 213, Amer. Math. Soc., Providence, R.I., 1979.

9. P. Nevai, Geza Freud, Orthogonal Polynomials and Christoffel Functions, A Case Study, J. Approx. Theory, 48(1986), 3-167.

10. G. Szegö, Orthogonal Polynomials, Amer. Math. Soc. Colloq. Pub., Vol. 23, Amer. Math. Soc., Providence, R.I., 1939,4th edn., 1975.

11. H. Widom, Polynomials Associated with Measures in the Complex Plane, J. Math. and Mech., 16(1967), 997-1013.

12. H. Widom, Extremal Polynomials Associated with a System of Curves in the Complex Plane, Adv. in Math., 3(1969), 127-232.

ASYMPTOTIC BEHAVIOR OF THE CHRISTOFFEL FUNCTION RELATED TO A CERTAIN UNBOUNDED SET

L.S. Luo and J. Nuttall
Department of Physics
The University of Western Ontario
London, Ontario, Canada. N6A 3K7

ABSTRACT. We study the asymptotic behavior, as the degree approaches infinity, of the Christoffel function at a fixed point z corresponding to a weight function of the type $\exp(-|z|^{\lambda})$ on the set $|\arg z| = \frac{\pi}{2} + \alpha$. The method generalizes that of Rakhmanov and also Mhaskar and Saff.

1. Introduction

In Quantum Mechanics there is a class of problems for which the required energy $E(\beta)$ may formally be expressed in terms of a parameter β by means of a power series with zero radius of convergence. Examples include the anharmonic oscillator and the hydrogen atom Stark effect. We have shown [2] that these series may be summed with the help of polynomials $Q_n(z)$ extremal with respect to the weight $\rho(z)\exp(-|z|^{\lambda})$, $\lambda \geq 1$, on the set $\Gamma(\alpha)$, where, with $-\frac{\pi}{2} < \alpha < \frac{\pi}{2}$,

$$(1.1) \qquad \Gamma(\alpha) = \{z \in \mathbb{C}: |\arg z| = \frac{\pi}{2} + \alpha\}.$$

We assume that $\rho(z)$ is a real positive locally integrable function for $z \in \Gamma(\alpha)$, that $\log \rho(z)$ is also locally integrable, and that

$$(1.2) \qquad \lim_{|z|\to\infty} |z|^{-\lambda} \log \rho(z) = 0.$$

We are here principally concerned with the quantity $K_n(z)$.

$$(1.3) \qquad K_n(z) = \sup_{\deg P_n \le n} \frac{|P_n(z)|^2}{\int_{\Gamma(\alpha)} |P_n(y)|^2 \rho(y) \exp(-|y|^\lambda) |dy|}.$$

$Q_n(z)$ is the polynomial (unique up to a multiplicative constant factor) for which the sup is attained. If $H_n(z)$ is the orthonormal polynomial of degree n corresponding to the weight, i.e.

$$(1.4) \qquad \int_{\Gamma(\alpha)} H_n(z)\, \overline{H_m(z)}\, \rho(z) \exp(-|z|^\lambda)\, |dz| = \delta_{nm},$$

then

$$(1.5) \qquad K_n(z) = \sum_{j=0}^{n} |H_j(z)|^2.$$

Thus $K_{n-1}(z)$ is the reciprocal of the Christoffel function as defined by Nevai [7] (see also Szegö [9]).

Here we analyze the asymptotics of $K_n(z)$ as $n \to \infty$ for fixed z with methods that closely follow those of Rakhmanov [8] and to a lesser extent Mhaskar and Saff [3,4]. We refer the reader to Rakhmanov's paper at many places where the argument is the same or is but slightly modified.

Our result is presented in the form of Theorem 4.5.

2. Minimum problem in potential theory

Suppose M_n is the set of all positive measures σ such that supp $\sigma \subset \Gamma(\alpha)$ and $\int_{\Gamma(\alpha)} d\sigma(t) = n$, and define the logarithmic potential

$$(2.1) \qquad V_\sigma(z) = -\int \log|z-t|\, d\sigma(t).$$

With $V_0(z) = |z|^\lambda$, $z \in \Gamma(\alpha)$, we define

$$(2.2) \qquad J(\sigma) = \int (V_\sigma(t) + V_0(t))\, d\sigma(t).$$

Mhaskar and Saff [4] have discussed the problem of minimizing $J(\sigma)$ for $\sigma \in M_n$ in the case when $\Gamma(\alpha)$ is replaced by the real line. As far as parts (a) to (f) of their Theorem 2.3 are concerned, their proofs apply with trivial modifications to our case $\Gamma(\alpha)$. In terms of the notation used here, patterned after Rakhmanov [8], we may therefore assert that there exists a unique measure $\upsilon \in M_n$ with compact support and of non-zero capacity such that

$$J(\upsilon) = j = \inf_{\sigma \in M_n} J(\sigma) > -\infty. \tag{2.3}$$

If we define

$$m = 2j - \int V_o(z)d\upsilon(z) \tag{2.4}$$

then

(2.5) $2V_\upsilon(z) + V_o(z) \geq m$, $z \in \Gamma(\alpha)$, except possibly for a set of capacity zero,

and

$$2V_\upsilon(z) + V_o(z) \leq m, \quad z \in \operatorname{supp} \upsilon. \tag{2.6}$$

From this information we have

<u>Lemma 2.1</u>

(2.7) (i) $\operatorname{supp} \upsilon = \Delta(R) = \{z \in \Gamma(\alpha) : |z| \leq R\}$, for some real positive R.

(2.8) (ii) $2V_\upsilon(z) + V_o(z) > m$, $z \in \Gamma(\alpha) \setminus \Delta(R)$.

Proof (i). The measure υ must be symmetric about the real axis for otherwise reflection in the real axis would lead to another measure with the same value of J, which would contradict uniqueness. The function $f(r)$ given by

$$f(r) = 2V_\upsilon(re^{i\alpha}) + V_o(re^{i\alpha}), \quad 0 \leq r < \infty, \tag{2.9}$$

is convex downwards, which may be seen by checking that

$$(2.10)\qquad -\log|re^{i\alpha}-se^{i\alpha}| - \log|re^{i\alpha}-se^{-i\alpha}|, \quad 0 \le r < \infty$$

is convex in r. The arguments of Rakhmanov [8] and Mhaskar and Saff [4] then show that supp υ must be connected so that (i) follows by symmetry.

(ii) If there were $r_0 > R$ for which $f(r_0) \le m$, then convexity shows that $f(r) < m$, $R < r < r_0$, which contradicts (2.5). □

3. Boundary value problem

We now outline the use of standard methods to find a real function $\tilde{V}(z)$, harmonic for $z \in \mathbb{C}\backslash\Delta(R)$, and continuous for $z \in \mathbb{C}$, which satisfies the conditions

$$(3.1)\qquad \tilde{V}(z) = 1/2(m-V_0(z)), \quad z \in \Delta(R)$$

$$(3.2)\qquad \tilde{V}(z) = -n\log|z| + O(|z|^{-1}), \quad |z| \to \infty.$$

With $\eta = 2\alpha/\pi$ it may be shown that there is a unique γ, $0 < \gamma < \pi$ for which

$$(3.3)\qquad \int_{-1}^{1} \frac{(t-c)(t-c^{-1})}{t^2}\left[\frac{t-1}{t+1}\right]^{\eta} dt = 0,$$

where the path of integration lies in the upper half-plane, and $c = e^{i\gamma}$. This can be seen by regarding (3.3) as an equation for $\cos\gamma$. We set

$$(3.4)\qquad z_0 = \int_{1}^{c} \frac{(t-c)(t-c^{-1})}{t^2}\left[\frac{t-1}{t+1}\right]^{\eta} dt$$

where the path of integration is on the unit circle in the upper half-plane, and define

$$(3.5)\qquad Z(w) = |z_0|^{-1}\int_{1}^{w} \frac{(t-c)(t-c^{-1})}{t^2}\left[\frac{t-1}{t+1}\right]^{\eta} dt, \quad |w| \ge 1$$

with integration path in $|t| \ge 1$.

<u>Lemma 3.1</u> The transformation $w \to z = Z(w)$ conformally maps $|w| > 1$ onto $z \in \mathbb{C}\setminus\Delta(1)$.

Proof. This is similar to a Schwarz transformation. [5, p.363]. □

Now write

(3.6) $$g(w) = -1/2|Z(w)|^{\lambda}, \quad |w| = 1$$

and define

(3.7) $$Q(w) = -(i\pi)^{-1} \int_{|t|=1} g(t)(t-w)^{-1}\, dt + Q_o$$

where

(3.8) $$Q_o = (2\pi i)^{-1} \int_{|t|=1} g(t)t^{-1}\, dt.$$

Then $Q(w)$ is analytic in $|w| > 1$ and satisfies [6, p.107]

(3.9) $$\operatorname{Re} Q(w) = g(w), \quad |w| = 1$$

where $Q(w)$ means the limit from $|w| > 1$.

<u>Lemma 3.2</u> The solution of (3.1), (3.2) is

(3.10) $$\tilde{V}(z) = \operatorname{Re}(R^{\lambda}Q(w) - n \log w + 1/2\ m),\ z = R\, Z(w),\ |w| \geq 1,$$

if and only if

(3.11) $$m = 2[n \log(|z_o|/R) - R^{\lambda}Q_o].$$

Proof. The function given is harmonic and satisfies (3.1). Since $Z(w) \sim |z_o|^{-1}w$ as $|w| \to \infty$, (3.2) will hold if (3.11) is true [1]

It must be possible to represent $\tilde{V}(z)$ as $V_{\mu}(z)$ for some signed measure μ with $\operatorname{supp} \mu \subset \Delta$. On $\Delta(R)$, $V_{\mu}(z) = V_{\nu}(z)$ except possibly for a set of capacity zero, and since $\Delta(R)$ has no irregular points, the uniqueness theorem shows that $\mu \equiv \nu$ [1, p.245], i.e.

$V_\upsilon(z) \equiv \tilde{V}(z)$. □

Now in the vicinity of the point c, for $|w| = 1$, $\arg Z(w) = \frac{\pi}{2} + \alpha$, so that we may write $|Z(w)| = Z(w)\exp(-i(\frac{\pi}{2} + \alpha))$, which shows that $g(w)$ has an analytic continuation in a neighborhood of $w = c$, as does $Q(w)$. We also have, with $B = be^{-2i\gamma}$, b real and positive,

$$(3.12) \qquad Z(w) = e^{i(\frac{\pi}{2}+\alpha)}(1 + B(w-c)^2) + O(|w-c|^3), \quad w \to c,$$

so that

$$\tilde{V}(z) - \tilde{V}(RZ(c)) = \mathrm{Re}[(R^\lambda Q'(c)-n/c)(w-c) + O(|w-c|^2)],\ w \to c$$

$$(3.13) \qquad = \mathrm{Re}\left[e^{i(\frac{\pi}{4} - \frac{\alpha}{2})}(BR)^{-1/2}(R^\lambda Q'(c)-n/c)(z-RZ(c))^{1/2}\right]$$

$$+ o(|z-RZ(c)|^{1/2}),$$

$$z \to RZ(c).$$

As in [8], we argue that the convexity of $V_\upsilon(z)$ outside $\Delta(R)$ and the need to satisfy (2.8) mean that

$$(3.14) \qquad R^\lambda = n/A, \quad A=cQ'(c).$$

Of course, this quantity must be real and positive, and we have found R.

<u>Lemma 3.3</u> It is possible to write $d\upsilon(z) = \upsilon'(z)|dz|$, $z\in\Delta(R)$, with

$$(3.15) \qquad \upsilon'(z) = \frac{n}{R}\, a\left[\frac{z}{R}\right], \quad z \in \Delta(R).$$

The function $|z|^{\frac{\eta}{1-\eta}} a(z)$ is continuous, $z \in \Delta(1)$.

Proof. The function $\upsilon'(z)$ is proportional [6] to the discontinuity of $\int_{\Delta(R)}(t-z)^{-1}d\upsilon(t)$ across $\Delta(R)$. With (3.10) this may be related to $Q'(w)$, which is analytic everywhere on $|w| = 1$ except at $w = \pm 1$. □

4. Bounds on $K_n(z)$

Following Rakhmanov [8], we first obtain in Lemma 4.3 bounds in the case $\rho \equiv 1$. The lower bound is constructed by evaluating (1.3) for a particular polynomial $P(z)$ of degree n, related to the measure υ^* defined below.

For x, y with $\arg x = \arg y = \frac{\pi}{2} + \alpha$, $|x| < |y|$, define

$$(4.1) \qquad \delta(x,y) = \{z \in \mathbb{C}: z = re^{i(\frac{\pi}{2}+\alpha)}, \ |x| \le r \le |y|\}.$$

Suppose that n is odd; the argument is similar for n even. Choose points z_j, $j = 1, \dots, (n+1)/2$ such that $\arg z_j = \frac{\pi}{2} + \alpha$, $|z_1| > |z_2| > \cdots > |z_{(n+1)/2}| = 0$, and

$$(4.2) \qquad \int_{\delta(z_1, RZ(c))} d\upsilon(t) = 1/2, \qquad \int_{\delta(z_{k+1}, z_k)} d\upsilon(t) = 1, \quad k = 1, \dots, (n-1)/2.$$

(Note that the symmetry of υ ensures that $z_{(n+1)/2} = 0$.) We define $z_k = \bar{z}_{n-k+1}$, $k = (n+1)/2, \dots, n$.

Define the measure υ^* as

$$(4.3) \qquad \upsilon^* = \sum_{j=1}^{n} \delta(z_j)$$

and

$$(4.4) \qquad V^*(z) = V_{\upsilon^*}(z) = -\sum_{j=1}^{n} \log|z - z_j|.$$

Lemma 4.1

(4.5) (i) $V_\upsilon(z) - V^*(z) \leq \left[\frac{1-\eta}{1-2\eta}\right] \log Cn, \quad z \in \mathbb{C},$

for some positive constant C.

(4.6) (ii) $V_\upsilon(z) - V^*(z) \geq -\log\left[1+R/D(z,\Delta(R))\right], \quad z \in \mathbb{C}\setminus\Delta(R),$

for fixed z and sufficiently large R. (The quantity $D(z,\Delta(R))$ means the distance from z to $\Delta(R)$.

Proof. (i) Suppose $x \in \Delta(R)$ lies in the interior of the interval $\delta(z_{k+1}, z_k)$, $k \leq (n-1)/2$. Then just as [8, p.171] we deduce that

(4.7) $$V(x)-V^*(x)\leq\log 2R - \int_{\delta(z_{k+1},z_k)} \log|x-t|\, d\upsilon(t).$$

We obtain the upper bound taking $k = (n-1)/2$, $x = 0$, and using Lemma 3.3.

(ii) The proof is as in [8, p.171]. ▫

One more piece of information is required before we can compute the lower bound. Define $W(z)$ by

(4.8) $$W(z) = \exp(m - 2V_\upsilon(z)).$$

Then we have

Lemma 4.2

(4.9) $$\lim_{n\to\infty} \frac{1}{2R} \int_{\Gamma(\alpha)} W(z)\exp(-|z|^\lambda)\,|dz| = 1.$$

Proof. Since $W(z)\exp[-|z|^\lambda] = 1$, $z \in \Delta(R)$, we have

(4.10) $$\int_{\Delta(R)} W(z)\exp(-|z|^\lambda)\,|dz| = 2R.$$

For $z \in \Gamma(\alpha)\setminus\Delta(R)$, we write

$$W(z)\exp(-|z|^\lambda) = \exp[2n\{\mathrm{Re}(\log w-Q(w)/A) - 1/2|Z(w)|^\lambda/A\}].$$

$$z = RZ(w).$$

$$(4.11) \qquad = \exp(2nh(w)) \quad \text{say.}$$

On account of Lemma 2.1 (ii) we know that $h(w) < 0$ for $z \in \Gamma(\alpha)\setminus\Delta(R)$. Thus $\int_{\Gamma(\alpha)\setminus\Delta(R)} W(z)\exp(-|z|^{\lambda})|dz|$ can be approximated for large n by the contribution from those z near $\Delta(R)$. For z near $RZ(c)$, i.e. w near c, we have

$$(4.12) \qquad h(w) = \text{Re}[C_1(w-c)^2] + O(|w-c|^3)$$

for some constant C_1. We conclude that

$$(4.13) \qquad \int_{\Gamma(\alpha)\setminus\Delta(R)} W(z)\exp(-|z|^{\lambda})|dz| = R(\text{const}.n^{-1} + O(n^{-2}))$$

and the Lemma follows. □

<u>Lemma 4.3</u> In the case $\rho \equiv 1$, for large n, and z in a fixed compact subset of $\mathbb{C}\setminus\Delta(R)$,

$$(4.14) \qquad |w|W(z)/\left[\pi D(z,\Delta(R))\right] \geq K_n(z)$$

$$\geq Cn^{-(3/\lambda+(2-2\eta)/(1-2\eta))}W(z),$$

$$z = RZ(w).$$

Proof. For the lower bound as [8, p.172] we use the polynomial $P(z) = e^{m/2} \prod_{i=1}^{n} (z-z_i)$, so that

$$(4.15) \qquad |P(z)| = \exp[1/2m - V^*(z)]$$

and

$$(4.16) \qquad |P(z)|^2 = W(z)\exp[2(V_{\upsilon}(z) - V^*(z))].$$

We use Lemmas 4.1 and 4.2 to estimate

$$(4.17)\qquad \frac{|P(z)|^2}{\int_{\Gamma(\alpha)} |P(z)|^2 \exp(-|z|^\lambda)\,|dz|}\,.$$

The proof of the upper bound is as in [8, p.160]. □

<u>Lemma 4.4</u> As $n \to \infty$ for z fixed, $|\arg z| \le \frac{\pi}{2}$, $|z| > 0$, we have

$$(4.18)\qquad \log W(z) = C_3 n^{1-\frac{1}{\lambda(\eta+1)}}\left[\operatorname{Re} z^{\frac{1}{\eta+1}}\right](1-A^{-1}Q'(1))(1+o(1))$$

where

$$(4.19)\qquad C_3 = 2A^{-\frac{1}{\lambda}}\left[|z_0|^{-1} 2^{-\eta}(\eta+1)^{-1}(1-c)(1-c^{-1})\right]^{-\frac{1}{\eta+1}}\,.$$

Proof. We have from (3.10), (4.8)

$$(4.20)\qquad \log W(z) = 2n\operatorname{Re}[\log w - A^{-1}Q(w)], \quad z/R = Z(w).$$

Near $w = 1$, (3.5) gives

$$(4.21)\qquad Z(w) = C_2(w-1)^{\eta+1}(1 + O(|w-1|)),$$

where

$$(4.22)\qquad C_2 = |z_0|^{-1} 2^{-\eta}(\eta+1)^{-1}(1-c)(1-c^{-1}),$$

so that, since $R \to \infty$,

$$(4.23)\qquad w-1 = (z/(RC_2))^{\frac{1}{\eta+1}}(1+O(|z/R|)).$$

Writing

$$(4.24)\qquad \log w - A^{-1}Q(w) = (1-A^{-1}Q'(1))(w-1) + o(|w-1|)$$

leads to the result. □

Theorem 4.5 Suppose $\rho(z)$, $z \in \Gamma(\alpha)$, is real, positive and integrable and $\log \rho(z)$ is locally integrable, and

$$(4.25) \qquad \lim_{|z|\to\infty} |z|^{-\lambda} \log \rho(z) = 0, \quad z \in \Gamma(\alpha).$$

With $K_n(z)$ defined by (1.3),

$$(4.26) \qquad \lim_{n\to\infty} \frac{\log K_n(z)}{n^{1-\frac{1}{\lambda(\eta+1)}}} = C_3(1-A^{-1}Q'(1)) \operatorname{Re}\left[z^{\frac{1}{\eta+1}}\right]$$

for fixed z, $|\arg z| \le \alpha$, $|z| > 0$, and $-\frac{\pi}{2} < \alpha < \frac{\pi}{2}$, $\lambda > 1$. For $\lambda = 1$, the result holds for $0 < \alpha < \frac{\pi}{2}$.

Proof. For $\rho(z) \equiv 1$ we use (4.18) in (4.14). The case $\rho(z) \not\equiv 1$ follows as in [8, Sec. 4].

We note that the above result in the case $\alpha = 0$, $\lambda > 1$, agrees with that of Rakhmanov [8].

REFERENCES

1. Landkof, N.S., (1972): Foundations of Modern Potential Theory. Berlin: Springer-Verlag.

2. Luo, L.S. and J. Nuttall, (1986): Approximation theory and calculation of energies from divergent perturbation series. Phys. Rev. Lett., 57, 2241-2243.

3. Mhaskar, H.N., E.B. Saff, (1984): Extremal problems for polynomials with exponential weights. Trans.Amer. Math. Soc. 285, 203-234.

4. Mhaskar, H.N., E.B. Saff, (1985): Where does the sup norm of a weighted polynomial live? Constr. Approx. 1: 71-91.

5. Moretti, G., (1964): Functions of a Complex Variable. Englewood Cliffs, N.J.: Prentice-Hall.

6. Muskhelishvili, N.I., (1953): Singular Integral Equations. Groningen: Noordhoff.

7. Nevai, P.: Geza Freud: Christoffel functions and orthogonal polynomials, J. Approx. Theory, 48, 3-167.

8. Rakhmanov, E.A., (1984): Asymptotic properties of polynomials orthogonal on the real axis. Math. U.S.S.R Sbornik, 47, 155-193.

9. Szegö, G., (1975): Orthogonal Polynomials, Amer. Math. Soc. Colloq. Pub, Vol. 23, Providence: American Mathematical Society.

SOME DISCREPANCY THEOREMS

by

H.N.Mhaskar
Department of Mathematics
California State University
Los Angeles, CA 90032,
USA

1. Introduction

In [5], E.B. Saff and I presented a general theory which unified many of the previously known results concerning incomplete polynomials as well as weighted polynomials on infinite intervals. A central aspect of this theory was a study of certain extremal problems involving a nonnegative weight function $w:R \to [0,\infty)$. We say that w is admissible if each of the following properties hold.

(W1) $\Sigma := \text{supp}\,(w)$ has positive capacity.
(W2) The restriction of w to Σ is continuous on Σ.
(W3) The set $Z := \{x \in \Sigma : w(x) = 0\}$ has capacity zero.
(W4) If Σ is unbounded, then $|x|w(x) \to 0$ as $|x| \to \infty$, $x \in \Sigma$.

Here, and in the sequel, the term capacity means inner logarithmic capacity (cf. [10,p.55]). For any set $E \subseteq R^2$, it's capacity will be denoted by $C(E)$. A property is said to hold q.e. (quasi-everywhere) on a set A if the subset E of A where it does not hold satisfies $C(E) = 0$. Let $\mathcal{P}_n$ denote the class of all polynomials of degree at most n and $\|\cdot\|_A$ denote the sup norm over a set A. In [5], we described the asymptotic behaviour of the errors in the weighted Chebyshev problem

---

† This research was funded in part, by an Affirmative Action Grant from The California State University, Los Angeles.

(1.1) $\quad E_n(w) := \inf\{\|[w(x)]^n\{x^n-p(x)\}\|_\Sigma : p \in \mathcal{P}_{n-1}\},\ n = 1,2,\ldots$

as well as the asymptotic properties (as $n \to \infty$) of the extremal polynomials $T_n(x,w) = x^n + \cdots \in \mathcal{P}_n$ which satisfy

(1.2) $\quad E_n(w) = \|[w(x)]^n T_n(x,w)\|_\Sigma$.

We summarize some of the results due to Saff and me in the following theorem.

Theorem 1.1: Let w be an admissible weight function,

(1.3) $\quad Q(x) := \log 1/w(x)$.

(a) There exists a finite constant $F := F(w)$ such that

(1.4) $\quad \lim\limits_{n\to\infty}[E_n(w)]^{1/n} = \exp\ (F)$.

(b) There exists a unit positive Borel measure μ_w such that $\mathcal{S}_w :=$ supp(μ_w) is compact, $\mathcal{S}_w \subseteq \Sigma\backslash Z$ and

(1.5) $\quad \int \log|x-t|d\mu_w(t) = Q(x) + F\quad$ q.e. on $\mathcal{S}_w$

(1.6) $\quad \int \log|x-t|d\mu_w(t) \le Q(x) + F\quad$ q.e. on Σ.

(c) Let $\mathcal{J} \subseteq R$ be a closed bounded interval containing $\mathcal{S}_w$ and $\{t_{k,n} : k = 1,\ldots,n\}$ be a triangular scheme of points lying in $\mathcal{J}$. With this scheme associate the sequence of polynomials

$$q_n(x) := \prod_{k=1}^{n} (x-t_{k,n}),\quad n = 1,2,\ldots ,$$

and a sequence of elementary unit measures $\{\nu^{(n)} : n = 1,2,\ldots\}$ where for every Borel set $\mathcal{B}$,

(1.7) $\quad \nu^{(n)}(\mathcal{B}) := \frac{1}{n}|\{k : t_{k,n} \in \mathcal{B}\}|,\quad n = 1,2,\ldots$.

Assume that

(1.8) $\lim_{n\to\infty} \sup \|[w(x)]^n q_n(x)\|^{1/n} \le \exp(F)$.

Then, in the weak-star topology,

(1.9) $\lim_{n\to\infty} \nu^{(n)} = \mu_w$.

In particular, (1.9) holds if $\{t_{k,n}\}$ are the zeros of $T_n(x,w)$.

We proved (1.4) only under certain additional conditions on w and hence, stated our results on the zeros of $T_n(x,w)$ only under these conditions. H. Stahl proved that (1.4) holds without any of these additional conditions. Important special cases include $w(x) := \exp(-|x|^\alpha)$, $\alpha > 0$, $\Sigma = \mathbb{R}$; $w(x) := x^\beta \exp(-x)$, $\beta > 0$, $\Sigma = [0,\infty)$ and $w(x) := (1-x)^\alpha(1+x)^\beta$, $0 < \alpha < 1-\beta < 1$, $\Sigma = [-1,1]$. In each of these cases, it was possible to compute explicitly the quantities F and μ_w. Moreover, $\mathcal{S}_w$ turned out to be an interval and μ_w was absolutely continuous. In the case when $w(x) = \exp(-|x|^\alpha)$, $\alpha > 1$, our computations were verified [7] independently by Rahmanov who also used the expression for μ_w to estimate the rate of convergence in (1.4). (He studied the case of the L^2-norm instead of the sup norm.)

In this paper, we give some estimates on the rate of convergence in (1.4) and (1.9) as well as a sufficient condition on w for μ_w to be absolutely continuous. Throughout this paper, we assume that Σ is an interval. Suppose that I is the smallest compact interval containing $\mathcal{S}_w$. Without loss of generality, we may assume that $I = [-1,1]$. We assume then that w is admissible and $Q(\cos\theta)$ is integrable on $[0,\pi]$. This will enable us to use results from the theory of Fourier series and H^p spaces. Our estimates will, in fact, be sharper when $\mathcal{S}_w = I$.

In the next section, we state our main results and prove them in Section 3.

2. Main Results

We adopt the following conventions. If $f:[-1,1] \to \mathbb{R}$, we put

$$\hat{f}(e^{i\theta}) := f(\cos\theta); \quad -\pi \le \theta \le \pi.$$

If $A \subseteq [-1,1]$ then $\hat{A} := \{e^{i\theta} : \cos\theta \in A;\ -\pi \le \theta \le \pi\}$. With the usual abuse of notation, we may also write $\hat{f}$ as a function on $[-\pi,\pi]$ and $\hat{A}$ as a subset of $[-\pi,\pi]$. Throughout this paper, c will denote a constant depending only on Q, the value of this constant may be different at different occurrences of the same symbol.

Theorem 2.1: We have

$$(2.1)\qquad \left|\frac{1}{n}\log E_n(w) - F\right| \le c\left\{\frac{\log n}{n} + \frac{1}{n}\sum_{k=1}^{n} \Omega(Q,\frac{\pi}{k})\right\}$$

where, for $\delta > 0$,

$$(2.2)\qquad \Omega(Q,\delta) := \sup_{\theta\in\mathcal{S}_w}\ \sup_{0<h\le\delta}\left\{\frac{1}{h}\int_0^h |\hat{Q}(\theta+t) + \hat{Q}(\theta-t) - 2\hat{Q}(\theta)|dt\right\}.$$

If $\mathcal{S} := \mathcal{S}_w$ is an interval (which is $[-1,1]$ because of our choice of normalization) then we may improve upon (2.1) and have

$$(2.3)\qquad \left|\frac{1}{n}\log E_n(w) - F\right| \le c\left\{\frac{\log n}{n} + \frac{1}{n}\sum_{k=1}^{n} \mathcal{E}_k\right\}$$

where

$$(2.4)\qquad \mathcal{E}_k := \min\{\|Q-P\|_{[-1,1]} : P \in \mathcal{P}_k\}.$$

In all the cases where explicit computations have been made so far, $\mathcal{S}_w$ is an interval and with the exception of the case when $w(x) := \exp(-|x|^\alpha)$, $0 < \alpha < 1$, $\Sigma = R$, Q satisfied a Lipschitz condition on this interval. The Jackson theorem on the degree of approximation along with (2.3) now yields

$$(2.5)\qquad \left|\frac{1}{n}\log E_n(w) - F\right| \le c\cdot\frac{\log n}{n}.$$

In the case when $w(x) = \exp(-|x|^\alpha)$, $0 < \alpha < 1$ and $\Sigma = R$, we get

$$(2.6)\qquad \left|\frac{1}{n}\log E_n(w) - F\right| \le \frac{c}{n^\alpha}.$$

Before we discuss the rate of convergence for the measures associated with a triangular scheme of points as in Theorem 1.1, we need to introduce additional notation. Any Borel measure ν on $[-1,1]$ gives rise to a Borel measure $\hat{\nu}$ on the unit circle (or $[-\pi,\pi]$) by means of the formula

$$(2.7)\qquad \hat{\nu}(\hat{B}) = 2\nu(B) \quad \text{for every Borel set } B \subseteq [-1,1].$$

In turn, this measure $\hat{\nu}$ corresponds to a function of bounded variation on $[-\pi,\infty)$ which will also be denoted by $\hat{\nu}$, by means of the formula

$$\hat{\nu}(t) := \int_{-\pi}^{t} d\hat{\nu}(u).$$

Following Ganelius [3] we measure the "niceness" of a measure by the signed modulus of continuity

$$(2.8)\qquad \omega(\nu,\delta) := \sup_{0\le h\le\delta}\ \sup_{t\in[-\pi,\pi]} \{\hat{\nu}(t+h) - \hat{\nu}(t)\}$$

and the "discrepancy" between two measures ν and σ by

$$(2.9)\qquad D(\nu,\sigma) := \sup_{[a,b]\subseteq[-\pi,\pi]} |\hat{\nu}([a,b]) - \hat{\sigma}([a,b])|.$$

Since $\delta\omega(\nu,\delta)$ is an increasing function of δ, we may define its inverse

$$(2.10)\qquad \mho(\nu,t) := \inf\{\delta : \delta\omega(\nu,\delta) \ge t\}.$$

We can now state our result concerning the rate of convergence of the measures in (1.9). This is an analogue of the discrepancy theorem of Erdös and Turan proved also by Ganelius [3].

<u>Theorem 2.2</u>: <u>Suppose that</u> $\mathcal{S} = [-1,1]$. <u>Let</u> $\{t_{k,n}\}$ <u>be a triangular scheme of points in</u> $[-1,1]$. <u>With this scheme we associate the sequence of polynomials</u> $\{q_n\}$ <u>and of measures</u> $\{\nu^{(n)}\}$ <u>as in Theorem 1.1(c). Let</u>

$$(2.11)\qquad e_n := \frac{1}{n}\log\|q_n(x)w(x)^n\|_{[-1,1]} - F, \quad n = 1,2,\ldots .$$

Then $\epsilon_n > 0$ for $n = 1,2,\ldots$, and

$$(2.12) \qquad D(\nu^{(n)},\mu_w) \le c\cdot\epsilon_n[\mho(\mu_w,\epsilon_n)]^{-1}.$$

In particular, (2.12) holds if $\{t_{k,n}\}$ are the zeros of $T_n(x;w)$.

If $\hat{\mu}_w$ is absolutely continuous and $\hat{\mu}_w' \in L^\infty[-\pi,\pi]$, then it is easy to see that $\mho(\mu_w,t) \sim t^{1/2}$. Lemma 3.1 will show that if $q_n(x) = T_n(x;w)$ then $\epsilon_n \le c(\log n)/n$. Thus, in this important special case,

$$(2.13) \qquad D(\nu^{(n)},\mu_w) \le c\{(\log n)/n\}^{1/2}.$$

In the following theorem, we give a sufficient condition on Q which guarantees that $\hat{\mu}_w$ is absolutely continuous and $\hat{\mu}_w' \in L^\infty[-\pi,\pi]$.

Theorem 2.3: Suppose that $\mathcal{S} = [-1,1]$, and $\hat{Q}$ is continuously differentiable on $[-\pi,\pi]$ and $\hat{Q}' \in \mathrm{Lip}\beta$ for some $\beta > 0$. Then $\hat{\mu}_w$ is absolutely continuous and $\hat{\mu}_w' \in C[-\pi,\pi]$.

We note that the conditions on Q in this theorem, stringent as they are, are in fact satisfied in all the cases when computations have been made. In the case when $w(x) = \exp(-|x|^\alpha)$, $0 < \alpha < 1$, $\Sigma = R$, these conditions are not satisfied, but then μ_w does not satisfy the conclusions of the theorem. It will be apparent from our proof that when $\mathcal{S} = [-1,1]$ and μ_w satisfies the conclusions of the theorem then $\hat{Q}$ is continuously differentiable.

Finally, we note that instead of considering the sup norm, we may consider any L^pnorm$(0 < p < \infty)$ in our definition of $E_n(w)$ in (1.2) and ϵ_n in (2.11). If w satisfies certain conditions under which we called it strongly admissible in [6], our results are substantially unchanged. The L^p version of these results follow immediately from the Nikolskii-type inequalities in [6]. For this reason, we have dealt exclusively with the sup norm.

3. Proof

In the proof of Theorem 2.1, we need the following lemma.

Lemma 3.1: Suppose that $f:[-1,1] \to [0,M]$ and

$$(3.1) \qquad \int_{-1}^{1} f(t)(1-t^2)^{-1/2}dt = 1.$$

Then for every integer $n = 2,3\ldots$, there exists a monic polynomial $P_n \in \mathscr{P}_{n-1}$ such that

$$(3.2) \qquad \frac{1}{n}\log|P_n(x)| \le \int_{-1}^{1} \log|x-t|f(t)(1-t^2)^{-1/2}dt + c\cdot\frac{\log(Mn)}{n}$$

where c is independent of f and n.

This lemma was proved in [4] when f is bounded from below; but we did not have the "correct" upper bound. This bound was obtained in [7], but under the condition that $f(t)(1-t^2)^{-1/2}$ be bounded. Our proof is an adaptation of the one in [7].

Proof of Lemma 3.1: Suppose that g is a measurable function such that $0 \le g(t) \le k$ and $\int_{-1}^{1} g(t)(1-t^2)^{-1/2}dt = 1$. Substituting $t =: \cos u$ and $x =: \cos\theta$, we get

$$\begin{aligned}
\mathscr{L}(x) &:= \int_{-1}^{1} \log|x-t|g(t)(1-t^2)^{-1/2}dt \\
&= \log 2 + \int_{-\pi}^{\pi} \log\left|\sin\left[\frac{\theta-u}{2}\right]\right|\hat{g}(u)du \\
&= \log 2 + \int_{-\pi}^{\pi} \log\left|\sin\frac{u}{2}\right|\hat{g}(\theta-u)du \\
&= \log 2 + \int_{0}^{\pi} \log\sin\frac{u}{2}\,[\hat{g}(\theta-u) + \hat{g}(\theta+u)]du.
\end{aligned}$$

Next, we write $h(u) := \hat{g}(\theta-u) + \hat{g}(\theta+u)$ and observe that, if $0 \le u \le \pi$, then $\log\sin\frac{u}{2} \ge \log\frac{u}{\pi}$. So,

$$(3.3) \qquad \mathscr{L}(x) \ge \log\left[\frac{2}{\pi}\right] + \int_{0}^{\pi} \log u\; h(u)du$$

$$= \log\left[\frac{1}{k\pi}\right] + \int_0^{\pi} \log(2ku)h(u)du$$

$$\geq \log(1/k\pi) + \int_0^{1/2k} \log(2ku)h(u)du$$

$$\geq \log(1/k\pi) + 2k \int_0^{1/2k} \log(2ku)du$$

$$= \log\left[\frac{2}{\pi}\right] + \log(1/2k) - 1$$

$$= -1 - \log(k\pi).$$

Now, we can proceed as in [7]. First, we find $-1 =: x_0 < x_1 < \cdots x_n = 1$ such that

$$\int_{x}^{x_{i+1}} f(t)(1-t^2)^{-1/2}dt = \frac{1}{n}, \quad i = 0,1,\ldots, n-1.$$

Next, we put $P_n(x) := \prod_{i=1}^{n-1} (x-x_i)$. If x is one of the x_i's then (3.2) is clear. Otherwise, we find an integer m such that $x_m < x < x_{m+1}$. Applying (3.3) with

$$g(t) := \begin{cases} nf(t); & x_m < t < x_{m+1} \\ 0 & \text{otherwise}, \end{cases}$$

we get

$$(3.4) \qquad \int_{x_m}^{x_{m+1}} \log|x-t| f(t)(2-t^2)^{-1/2}dt \geq \frac{-1-2\log(Mn)}{n}$$

$$\geq -c\cdot\frac{\log(Mn)}{n}.$$

So,

$$(3.5) \qquad \int_{-1}^{1} \log|x-t| f(t)(1-t^2)^{-1/2}dt$$

$$= \sum_{i=0}^{m-1} \int_{x_i}^{x_i} \log|x-t| f(t)(1-t^2)^{-1/2} dt$$

$$+ \int_{x_m}^{x_{m+1}} \log|x-t| f(t)(1-t^2)^{-1/2} dt$$

$$+ \sum_{i=m+1}^{n-1} \int_{x_i}^{x_{i+1}} \log|x-t| f(t)(1-t^2)^{-1/2} dt$$

$$\geq \frac{1}{n} \sum_{i=0}^{m-1} \log|x-x_{i+1}| - c\cdot\frac{\log(Mn)}{n}$$

$$+ \frac{1}{n} \sum_{i=m+1}^{n-1} \log|x-x_i|$$

$$= \frac{1}{n} \log|P_n(x)| - c\cdot\frac{\log(Mn)}{n} . \qquad \square$$

When $\hat{\mu}_w$ is absolutely continuous and $\hat{\mu}_w' \in L^\infty[-\pi,\pi]$, this lemma immediately implies that (cf. the proof of Theorem 4.2 of [5])

(3.6) $$\left|\frac{1}{n} \log E_n(w) - F\right| \leq c\cdot\frac{\log n}{n} .$$

In the general case, we may use smoothing, for instance by the Fejer sums. We put

(3.7) $$L(t) := \log \sin\left|\frac{t}{2}\right|, \quad -\pi \leq t \leq \pi$$

(3.8) $$F_n(t) := \frac{1}{2\pi n}\left[\frac{\sin nt/2}{\sin t/2}\right]^2, \quad -\pi \leq t \leq \pi$$

and let $*$ denote the convolution operation.

<u>Proof of Theorem 2.1</u>: In view of (1.6), we see that

(3.9) $$(L * d\hat{\mu}_w)(\theta) \leq \hat{Q}(\theta) + F \quad \text{q.e} \quad \text{on} \quad [-\pi,\pi],$$

and hence,

$$(3.10)\qquad L*(F_n*d\hat{\mu}_w)(\theta) = F_n*(L*d\hat{\mu}_w)(\theta) \le (F_n*\hat{Q})(\theta)+F,\ -\pi \le \theta \le \pi.$$

Now, $F_n * d\hat{\mu}_w$ is an even trigonometric polynomial. So, there is an algebraic polynomial f_n such that

$$(3.11)\qquad f_n(\cos\theta) = \hat{f}_n(\theta) = (F_n*d\hat{\mu}_w)(\theta);\quad -\pi \le \theta \le \pi.$$

Moreover, it is clear that

$$(3.12)\qquad 0 \le f_n(x) \le c\cdot n,\quad -1 \le x \le 1,$$

and that

$$(3.13)\qquad (L*\hat{f}_n)(\theta) = \int_{-1}^{1} \log|\cos\theta - t| f_n(t)(1-t^2)^{-1/2}dt.$$

Hence, Lemma 3.1 gives a monic polynomial $P_n \in \mathcal{P}_n$ such that for $x = \cos\theta$,

$$(3.14)\qquad \frac{1}{n}\log|P_n(x)| \le (L*\hat{f}_n)(\theta) + c\cdot\frac{\log n}{n}$$

$$\le (F_n*\hat{Q})(\theta) + c\cdot\frac{\log n}{n} + F$$

$$= Q(x) + F + c\cdot\frac{\log n}{n} + (F_n*\hat{Q})(\theta)-\hat{Q}(\theta).$$

In particular, taking sup over $x \in \mathcal{S}$ (i.e. $\theta \in \hat{\mathcal{S}}$), we get

$$(3.15)\qquad \frac{1}{n}\log\|P_n(x)w(x)^n\|_\Sigma - F \le c\cdot\frac{\log n}{n} + \sup_{\theta\in\hat{\mathcal{S}}}|F_n*\hat{Q}(\theta)-\hat{Q}(\theta)|.$$

Now, in view of Theorems 2.1 (c') and 2.1 (d) of [5],

$$(3.16)\qquad \|P_n w^n\|_{\mathcal{S}} = \|P_n w^n\|_\Sigma \ge E_n(w) \ge \exp(nF).$$

Hence (3.15) and standard estimates on the approximation by Fejer sums [1], [8] yield (2.1); and when $\mathcal{S} = [-1,1]$ also (2.3). □

Theorem 2.2 is an application of the following theorem of Ganelius.

<u>Lemma 3.2 ([3]): Let V be a real periodic function of bounded variation and let k be a real periodic integrable function with the Fourier coefficients</u>

$$k_m := (2\pi)^{-1} \int_{-\pi}^{\pi} e^{-im\tau} k(\tau)\, d\tau .$$

<u>Suppose</u> $k_m \neq 0$ <u>if</u> $m \neq 0$. <u>Then</u>

$$\sup_t |V(t) - V_0| \leq c\left\{ \|k * dV\|_1 \sum_{m=1}^{n-1} |mk_m|^{-1} + \omega(V, \tfrac{1}{n}) \right\} \tag{3.17}$$

<u>where</u>

$$V_0 := \frac{1}{2\pi} \int_{-\pi}^{\pi} V(t)\, dt .$$

Ganelius applied this theorem with the difference of a measure and the arc length measure on the circle for dV and

$$L_\rho(t) := \log|1 - \rho e^{it}|, \quad 0 < \rho < 1 \tag{3.18}$$

for $k(t)$ to obtain a simple proof of a discrepancy theorem of Erdös and Turan.

<u>Proof of Theorem 2.2</u>: Let

$$V(t) := \hat{\mu}_w(t) - \hat{\nu}^{(n)}(t). \tag{3.19}$$

Then $V_0 = 0$ and for each $\rho \in (0,1)$

$$D(\nu^{(n)}, \mu_w) \leq 2 \sup_t |V(t)| = 2 \sup_t |V(t) - V_0| \tag{3.20}$$

$$\leq c\left\{ \|L_\rho * dV\|_1 \cdot \sum_{m=0}^{N-1} \rho^{-m} + \omega(V, \tfrac{1}{N}) \right\} .$$

But

(3.21) $\omega(V, \frac{1}{n}) \le \omega(\mu_w, \frac{1}{N})$.

Moreover,

$$\int_{-\pi}^{\pi} (L_\rho * dV)(\theta) d\theta = \int_{-\pi}^{\pi} \int_{-\pi}^{\pi} \log|1-\rho e^{i(\theta-\tau)}| dV(\tau) d\theta$$

$$= \int_{-\pi}^{\pi} \int_{-\pi}^{\pi} \log|1-\rho e^{i(\theta-\tau)}| d\theta \, dV(\tau) = 0, \quad 0 < \rho < 1.$$

Hence as in [3],

(3.22) $\|L_\rho * dV\|_1 \le 2|\inf_\theta (L_\rho * (dV))(\theta)| \le 2|\inf_\theta (L * dV)(\theta)|$.

(Here we also used the fact that $\int dV = 0$.) Now, if $x := \cos\theta$, then

$$(L * dV)(\theta) = \int \log|x-t| d\mu_w(t) - \int \log|x-t| d\nu^{(n)}(t)$$

$$= -\frac{1}{n} \log|q_m(x) w(x)^n| + F.$$

Thus,

(3.23) $\inf_\theta (L * dV)(\theta) = -\frac{1}{n} \log\|q_m(x) w(x)^n\| + F = -\epsilon_n$.

We substitute from (3.22), (3.21) and (3.20) into (3.19) and choose N so that

$$N w(\mu_w, \frac{1}{N}) = \epsilon_n.$$

This completes the proof of (2.12). □

Theorem 2.3 will follow from the following Lemma.

<u>Lemma 3.3</u>: <u>Let</u> $\mathcal{S} = [-1, 1]$ <u>and set</u>

(3.24) $h(z) := \frac{1}{2\pi} \int_{-\pi}^{\pi} \hat{Q}(\theta) \left[\frac{1+e^{i\theta}z}{1-e^{i\theta}z}\right] d\theta, \quad |z| < 1.$

<u>Then</u>

(3.25) $\operatorname{Re} h(e^{i\theta}) = \hat{Q}(\theta), \quad -\pi \le \theta \le \pi$

(3.26) $$h(z) = \int_{-\pi}^{\pi} \log(1-ze^{i\tau})d\hat{\mu}_w(\tau) + \frac{1}{2\pi}\int_{-\pi}^{\pi} \hat{Q}(\theta)d\theta \qquad (|z|<1).$$

<u>Moreover, if</u> $\hat{Q}$ <u>satisfies the conditions of Theorem 2.3 then</u> h' <u>is continuous on</u> $|z| \le 1$.

<u>Proof</u>: Formula (3.24) follows from the fact that $\operatorname{Re} h(z)$ is the Poisson integral of the continuous function $\hat{Q}$. Since $\mathcal{S} := [-1,1]$, the quantity F is given by (cf. Example 3,[5]),

(3.27) $$F = \log\frac{1}{2} - \frac{1}{2\pi}\int_{-\pi}^{\pi}\hat{Q}(\theta)d\theta.$$

Moreover (1.5) holds everywhere on $[-1,1]$ (cf. the remark following the proof of Lemma 4.3 of [5]). So,

(3.28) $$\operatorname{Re} \int \log(1-ze^{it})d\hat{\mu}_w(t) + \frac{1}{2\pi}\int_{-\pi}^{\pi}\hat{Q}(t)dt = \hat{Q}(e^{i\theta}),$$

$$z = e^{i\theta}, \quad -\pi \le \theta \le \pi.$$

Denoting the right hand side of (3.25) by $\Lambda(z)$, we thus see that Λ and h are both analytic on $|z| < 1$, $\Lambda(0) = h(0)$, $\operatorname{Re}\Lambda$ and $\operatorname{Re} h$ are continuous on $|z| \le 1$ (Lemma 4.3 of [5]) and equal on $|z| = 1$. This proves (3.25). Next, if f is a 2π - periodic continuous function, we put

(3.29) $E_n^*(f) := \inf\|f-T\|$

where the inf is over all trigonometric polynomialsof order at most n. If $\hat{Q}$ is continuously differentiable and $\hat{Q}' \in \mathrm{Lip}\beta$ then (cf.5.1.4 (18) of [9])

$$E_n^*(\hat{Q}) = O\left(\frac{1}{n^{1+\beta}}\right).$$

Then (cf. Theorem 6.1.4 of [9]), $\hat{R}$ is also continuously differentiable and $\hat{R} \in \mathrm{Lip}\beta$. Then the Lemma is proved in view of Theorem 3.1.1 of [2]. □

Proof of Theorem 2.3: Differentiating under the integral sign in (3.25), we get

$$h'(z) = \int_{-\pi}^{\pi} \frac{e^{i\tau}}{ze^{i\tau}-1} d\hat{\mu}_w(\tau), \quad |z| < 1.$$

Recalling that $\int d\hat{\mu}_w = 2$, we see that

$$(3.30) \qquad 1 - zh'(z) = \frac{1}{2}\int_{-\pi}^{\pi} \frac{1+z\cdot e^{i\tau}}{1-z\cdot e^{i\tau}} d\hat{\mu}_w(\tau), \quad |z| < 1.$$

Thus, $\mathrm{Re}(1-zh'(z))$, which is continuous on $|z| \le 1$, in view of Lemma 3.3, is the Poisson integral of $d\hat{\mu}_w$. This completes the proof. □

Remark: If $\hat{\mu}_w$ is absolutely continuous, (3.29) shows that $izh'(z)$ is continuous on $|z| \le 1$. In particular, $\hat{Q}'$ exists and is continuous.

REFERENCES

1. S. Aljanic, R. Bojanic and M. Tomic, On the degree of convergence of Fejer-Lebesgue sums, L'Ensignment Math., 15 (1969), 21-28.

2. P.L. Duren, Theory of H^p spaces, Academic Press, New York, 1970.

3. T. Ganelius, Some applications on a lemma of Fourier series, Publ. de L'Institute Math., 11(1957), 9-18.

4. H.N. Mhaskar and E.B. Saff, Extremal problems for polynomials with exponential weights, Trans. Amer. Math Soc., 285(1984), 203-234.

5. ____________, Where does the sup norm of a weighted polynomial live? (A generalization of incomplete polynomials), Constr. Approx., 1(1985), 71-91.

6. ____________, Where does the L^p-norm of a weighted polynomial live? To appear.

7. E.A. Rakhmanov, Asymptotic properties of polynomials orthogonal on the real axis, Matem. Sb. 119(61) (1982), Math. USSR Sbornik, 47(1) (1984), 155-193.

8. S.B. Stechkin, The approximation of periodic functions by Fejer

9. A.F. Timan, Theory of Approximation of functions of a real variable, MacMillan, New York, 1963.

10. M. Tsuji, Potential Theory in Modern Function Theory, Maruzen Tokyo, 1959.

Properties of Projections Obtained By Averaging Certain Polynomial Interpolants

Judith Palagallo-Price and Thomas E. Price

Department of Mathematical Sciences

The University of Akron

Akron, Ohio 44325

Abstract. We describe a way to compute polynomial approximants to analytic functions $f(z)$ in the unit disk by forming the average of m polynomials of degree $n-1$, each of which interpolates $f(z)$ at n equidistant points on the unit circle. The paper discusses properties of the projections so defined. Norms of these projections are calculated and the asymptotic behavior is characterized. Furthermore, these averages are used to approximate Laurent sections.

1. Introduction

In a linear space X equipped with the norm $\|\cdot\|$, let there be defined a subspace Y and a map $M : X \mapsto Y$. For $x \in X$, Mx is said to be an approximation to x from Y with error $E := \|x - Mx\|$. Assume that for $x \in X$ there is a $y_x \in Y$ such that $\|x - y_x\| \leq \|x - y\|$ for all $y \in Y$; y_x is then referred to as a best approximation to x from Y. The approximation Mx is said to be near-best within a relative distance ρ if $\|x - Mx\| \leq (1+\rho)\|x - y_x\|$. Following Geddes and Mason [10], we shall refer to Mx as a practical near-best approximation if the factor ρ is "acceptably small". In particular, if $\rho \leq 9$, then no more than one decimal place of accuracy is lost in taking Mx in place of y_x.

An important class of approximation maps is that of projections. A projection $P : X \mapsto Y$ is a map which is bounded, linear and idempotent. Furthermore, a projection P satisfies the inequality

$$\|x - Px\| \leq \|I - P\| \inf\{\|x - y\| : y \in Y\},$$

where I denotes the identity operator on X. (Here and in what follows, operator norms are assumed to be those induced by the linear space norm.) If to each $x \in X$ there corresponds a best approximation $y_x \in Y$, the above inequality implies that

$$\|x - Px\| \leq (1 + \|P\|)\, \|x - y_x\|.$$

Thus, a projection P produces a near-best approximation within a relative distance $\|P\|$. (A review of the usefulness of such approximations can be found in [10].)

In that which follows we shall need the

<u>**Definition.**</u> Let Σ be a prescribed family of projections all having the same domain X and range Y. If $P_0 \in \Sigma$ is such that $\|P_0\| \leq \|P\|$ for all $P \in \Sigma$, then P_0 is minimal in Σ.

2. Preliminaries and Review of Literature

Let $\gamma_\rho := \{z : |z| = \rho, \quad \rho > 0\}$ and denote the space of all functions f continuous on γ_ρ by $C(\gamma_\rho)$. Let $A(\gamma_\rho)$ be the space of functions $f \in C(\gamma_\rho)$ which are analytic on $D := \{z : |z| < \rho\}$. For the case $\rho = 1$ we simply write γ. Equip $C(\gamma)$ with the Tchebychev (minimax) norm:

$$\|f\| := \sup\{|f(z)| : z \in \gamma\}, \quad f \in C(\gamma).$$

Let Π_{n-1} denote the space of complex algebraic polynomials of degree at most $n-1$. For $f \in C(\gamma)$ and n a positive integer, let $L_{n-1}(z; f) \in \Pi_{n-1}$ denote the unique polynomial interpolant of f in the n-th roots of unity; i.e., the nodes ω^j, $j = 0, \dots, n-1$, where $\omega := e^{2\pi i/n}$. Finally, define the discrete Fourier projection $F_{n-1} : C(\gamma) \mapsto \Pi_{n-1}$ by

$$(F_{n-1} f)(z) := L_{n-1}(z; f). \tag{2.1}$$

In [10] it was shown that this projection is a practical near-best approximation to f since $\|F_{n-1}\| \leq 5.4$ for $n \leq 1000$. This result depends on the following interesting inequality due to Gronwall [11]:

$$\|F_{n-1}\| \leq \frac{1}{n} \sum_{\nu=0}^{\infty} \csc\left(\frac{2\nu + 1}{2n}\right)\pi. \tag{2.2}$$

More recently, Erdős [6] conjectured that F_{n-1} is minimal in the family of all projections from $C(\gamma)$ into Π_{n-1} induced by interpolation at n distinct points on γ. This conjecture was shown to be true by Brutman [2] and Brutman and Pinkus [3].

As we shall see, for $f \in A(\gamma)$, the discrete Fourier projection of f into Π_{n-1} is closely related to the Taylor section of degree $n-1$. For this reason we mention a few related results concerning these sections. Given $f \in A(\gamma)$, $f(z) := \sum_{s=0}^{\infty} a_s z^s$, let $P_{n-1}(z; f) := \sum_{s=0}^{n-1} a_s z^s$ be its $(n-1)$-st degree Taylor polynomial. The associated projection $T_{n-1} : A(\gamma) \mapsto \Pi_{n-1}$, defined by $(T_{n-1}f)(z) := P_{n-1}(z; f)$, has been shown ([10]) to be minimal in the family of all projections from $A(\gamma)$ into Π_{n-1}. Landau obtained the following exact expression for $\|T_{n-1}\|$:

$$\|T_{n-1}\| \equiv G_{n-1} := 1 + \left(\frac{1}{2}\right)^2 + \left(\frac{1\cdot 3}{2\cdot 4}\right)^2 + \cdots + \left(\frac{1\cdot 3\cdots(2n-3)}{2\cdot 4\cdots 2(n-1)}\right)^2.$$

He also proved that $G_{n-1} \sim \frac{1}{\pi}\log(n-1)$. These and related results can be found in [5]. An investigation of G_{n-1} was continued by Watson [17] who proved that

$$G_{n-1} < \tau_{n-1} := \frac{2}{\pi}\int_0^{\pi/2} \frac{|\sin n\theta|}{\sin\theta}\, d\theta, \tag{2.3}$$

where τ_{n-1} is the $(n-1)$-st classical Lebesgue constant. Brutman [1] recently obtained the more restrictive bounds

$$1 + \frac{1}{\pi}\log n \le \|T_{n-1}\| = G_{n-1} \le 1.0663 + \frac{1}{\pi}\log n. \tag{2.4}$$

Since this establishes that $\|T_{n-1}\| \le 3.3$ for $n \le 1000$, it follows that $T_{n-1}f$ is a practical near-best approximation to $f \in A(\gamma)$. Geddes and Mason [10] presented efficient algorithms for computing the two projections T_{n-1} and F_{n-1} based on the fast Fourier transform.

More recently, Palagallo and Price [15] considered projections obtained by averaging certain polynomial interpolants. Specifically, let $(R_\phi f)(z) := f(e^{i\phi}z)$. Then $L_{n-1}^{(\phi)} := R_{-\phi}F_{n-1}R_\phi : C(\gamma) \mapsto \Pi_{n-1}$ is the projection obtained by interpolation to $f(z)$ in the nodes $\{e^{i\phi}\omega^\nu\}_{\nu=0}^{n-1}$. For m a positive integer, we let $\phi_\mu := \frac{2\pi}{mn}\mu$, $\mu = 0, \ldots, m-1$, and define the mean projection

$$M_{n-1,m} := \frac{1}{m}\sum_{\mu=0}^{m-1} L_{n-1}^{(\phi_\mu)}. \tag{2.5}$$

Corresponding to $M_{n-1,m}$ is the polynomial in Π_{n-1} given by

$$A_{n-1,m}(z; f) := (M_{n-1,m}f)(z), \qquad f \in C(\gamma). \tag{2.6}$$

Note that $A_{n-1,m}(z;f)$ is the average of polynomial interpolants, each of which has as its nodes a rotation of the n-th roots of unity. Indeed, the μ-th polynomial $L_{n-1}^{(\phi_\mu)}f$, $0 \leq \mu \leq m-1$, interpolates f in the nodes $\exp(i\phi_\mu)\,\omega^\nu = e^{[2\pi/mn](\mu+\nu m)i}$, $\nu = 0,\ldots,n-1$. Note that these nodes collectively form the mn-th roots of unity.

Since T_{n-1} is a minimal projection and $M_{n-1,m}$ may be viewed as the average of interpolants each having norm $\|F_{n-1}\|$, we have $\|T_{n-1}\| \leq \|M_{n-1,m}\| \leq \|F_{n-1}\|$. Hence $M_{n-1,m}$ is also a practical near-best approximation. Furthermore, the operator $M_{n-1,m}$ is defined for any function whose domain contains the mn-th roots of unity.

It is with this averaging technique that we concern ourselves in this paper. A similar averaging technique restricted to trigonometric polynomial interpolants was discussed by Morris and Cheney in [14]. Since F_{n-1} is easily computed using the Fast Fourier Transform, it is not surprising that this transform can also be used to compute $A_{n-1,m}$. It can be shown that this procedure decreases the computational effort required to compute certain least squares approximations. A discussion of this topic is reserved for a later paper.

This report is organized as follows. In Section 3 there is a discussion of some properties of $M_{n-1,m}$. A method for computing $\|M_{n-1,m}\|$ was given in [15], but for completeness, a somewhat revised approach is presented in Section 4. In Section 5 asymptotic results concerning $\|M_{n-1,m}\|$ are developed. In Section 6 we apply the averaging technique to approximate Laurent series.

3. THE MEAN PROJECTION

We begin this section with an interesting representation for $M_{n-1,m}$.

Proposition 3.1. The mean projection satisfies

$$M_{n-1,m} = T_{n-1}F_{mn-1}.$$

This property was noted by Rivlin [10] for functions in $A(\gamma_\rho)$, $\rho > 1$. His proof used the Taylor coefficients for f. The one given below is more general in that it does not assume that f is analytic in any circle containing the origin. In what follows we will use the notation $\varsigma \equiv \varsigma(m,n) := \exp(2\pi i/mn)$.

Proof of Proposition 3.1. Let f be defined on γ. It suffices to show that $A_{n-1,m}(z;f) =$

$P_{n-1}(z; L_{mn-1}(\cdot; f))$. To this end note that one can easily verify the equalities

$$L_{mn-1}(z; f) = \frac{z^{mn}-1}{mn} \sum_{j=0}^{mn-1} \frac{\varsigma^j f(\varsigma^j)}{z-\varsigma^j}$$

$$= \frac{1}{mn} \sum_{j=0}^{mn-1} \sum_{k=0}^{mn-1} \varsigma^{j(mn-k)} f(\varsigma^j) z^k.$$

Consequently,

$$P_{n-1}(z; L_{mn-1}(\cdot; f)) = \frac{1}{mn} \sum_{j=0}^{mn-1} \sum_{k=0}^{n-1} \varsigma^{j(mn-k)} f(\varsigma^j) z^k$$

$$= \frac{1}{mn} \sum_{j=0}^{mn-1} \varsigma^{-j(n-1)} f(\varsigma^j) \frac{z^n - \varsigma^{jn}}{z-\varsigma^j}.$$

For $0 \le j \le mn-1$, j has the unique representation $j = \mu + \nu m$, $0 \le \nu \le n-1$ and $0 \le \mu \le m-1$. So, the last equation becomes

$$\frac{1}{mn} \sum_{\mu=0}^{m-1} \sum_{\nu=0}^{n-1} \frac{\varsigma^{\mu+\nu m} f(\varsigma^{\mu+\nu m})}{\varsigma^{\mu n}} \cdot \frac{z^n - \varsigma^{\mu n}}{z - \varsigma^{\mu+\nu m}}$$

$$= \frac{1}{m} \sum_{\mu=0}^{m-1} L_{n-1}^{(\phi_\mu)}(z; f) = A_{n-1,m}(z; f),$$

and this completes the proof. ∎

In view of Proposition 3.1, one would suspect that $M_{n-1,m}f$ is a reasonable approximation to $T_{n-1}f$ for $f \in A(\gamma)$. This is indeed the case as shown by the next proposition.

Proposition 3.2. For $f \in A(\gamma)$, $\lim_{m\to\infty} M_{n-1,m}f = T_{n-1}f$. That is, the mean projection converges pointwise to the Taylor projection.

A complete proof of Proposition 3.2 is quite lengthy and appears in [15]. We provide the following alternative and concise proof using the more stringent hypothesis $f \in A(\gamma_\rho)$, $\rho > 1$.

Proof of Proposition 3.2. Let $f \in A(\gamma_\rho)$, $\rho > 1$, with Taylor series $f(z) := \sum_{s=0}^{\infty} a_s z^s$. Then

$$f(z) = \sum_{\nu=0}^{\infty} \sum_{\mu=0}^{mn-1} a_{\mu+\nu mn} z^{\mu+\nu mn}$$

for $|z| \le 1$. Thus, $(L_{mn-1}f)(z) = \sum_{\nu=0}^{\infty}\sum_{\mu=0}^{mn-1} a_{\mu+\nu mn}\, z^{\mu}$ since $(\varsigma^j)^{\mu+\nu mn} = (\varsigma^j)^{\mu}$, $0 \le j \le mn-1$. Then, by Proposition 3.1,

$$\begin{aligned} A_{n-1,m}(z;f) &= P_{n-1}\left(z; L_{mn-1}(\cdot;f)\right) \\ &= \sum_{\nu=0}^{\infty}\sum_{\mu=0}^{n-1} a_{\mu+\nu mn}\, z^{\mu} \\ &= (T_{n-1}f)(z) + \sum_{\nu=1}^{\infty}\sum_{\mu=0}^{n-1} a_{\mu+\nu mn}\, z^{\mu}. \end{aligned}$$

So, for $|z| \le 1$,

$$|T_{n-1}(z;f) - A_{n-1,m}(z;f)| \le \sum_{\nu=1}^{\infty}\sum_{\mu=0}^{n-1} |a_{\mu+\nu mn}| = O\left(\rho^{-mn}\right). \blacksquare$$

The above proof actually establishes the stronger result $\lim_{m\to\infty} M_{n-1,m} = T_{n-1}$ on the space $A(\gamma_\rho)$, $\rho > 1$. This operator convergence does not hold, however, on $A(\gamma)$. To see this we note that later we shall establish $\lim_{m\to\infty} \|M_{n-1,m}\| = \tau_{n-1}$, where τ_{n-1} was defined in (2.3). This, along with the inequality in (2.3) establishes that

$$\lim_{m\to\infty} \|M_{n-1,m}\| > \|T_{n-1}\|.$$

Proposition 3.3. Let f be defined on $\{\varsigma^j\}_{j=0}^{mn-1}$. Then $A_{n-1,m}(z;f)$ is the best least squares approximation to f over the mn-th roots of unity.

This was proved by Rivlin [16] for $f \in A(\gamma_\rho)$, $\rho > 1$, using the Taylor coefficients of f. We shall establish the general result.

Proof of Proposition 3.3. Recall the proof of Proposition 3.1 and consider

$$\begin{aligned} f(\varsigma^{\iota}) - A_{n-1,m}(\varsigma^{\iota};f) &= f(\varsigma^{\iota}) - P_{n-1}\left(\varsigma^{\iota}, L_{mn-1}(\cdot;f)\right) \\ &= f(\varsigma^{\iota}) - \frac{1}{mn}\sum_{j=0}^{mn-1}\sum_{k=0}^{n-1} \varsigma^{j(mn-k)} (\varsigma^{\iota})^k f(\varsigma^j) \\ &= f(\varsigma^{\iota}) - \frac{1}{mn}\sum_{j=0}^{mn-1}\sum_{\mu=0}^{n-1} \varsigma^{\iota k - jk} f(\varsigma^j) \end{aligned}$$

$$= f(\varsigma^{\iota}) - L_{mn-1}(f;\varsigma^{\iota}) + \frac{1}{mn}\sum_{j=0}^{mn-1}\sum_{k=n}^{mn-1}\varsigma^{\iota k-jk}f(\varsigma^{j}).$$

So, for $s = 0,\ldots,n-1$, the discrete inner products

$$\begin{aligned}\langle f - A_{n-1,m}(\cdot;f), z^{s}\rangle &= \sum_{\iota=0}^{mn-1}\left[f(\varsigma^{\iota}) - A_{n-1,m}(\varsigma^{\iota};f)\right]\varsigma^{-\iota s}\\ &= \frac{1}{mn}\sum_{\iota=0}^{mn-1}\sum_{j=0}^{mn-1}\sum_{k=n}^{mn-1}\varsigma^{\iota k-jk}\varsigma^{-\iota s}f(\varsigma^{j})\\ &= \frac{1}{mn}\sum_{j=0}^{mn-1}f(\varsigma^{j})\sum_{k=n}^{mn-1}\varsigma^{-jk}\sum_{\iota=0}^{mn-1}\varsigma^{\iota(k-s)}.\end{aligned}$$

The sum over ι in the above is zero since $(k-s)$ is not an integer multiple of mn. This completes the proof. ∎

4. THE NORM OF THE MEAN PROJECTION

In this section we outline a method for computing the norm of the projection $M_{n-1,m}$. A more detailed analysis can be found in [15].

Theorem 1. Let $M_{n-1,m}$ denote the mean projection. Then

(4.1) $$\|M_{n-1,m}\| = \tau_{n-1}^{(m)}, \qquad m = 1,2,\ldots,$$

where

(4.2) $$\tau_{n-1}^{(m)} := \begin{cases} \dfrac{2}{mn}\displaystyle\sum_{\nu=0}^{t-1}\frac{\cot\dfrac{(2\nu+1)\pi}{2n}}{\sin\left(\dfrac{1}{2m}-\dfrac{2\nu+1}{2mn}\right)\pi}, & \text{if } n = 2t\\[3ex] \dfrac{2}{mn}\displaystyle\sum_{\nu=0}^{t-1}\frac{\cot\dfrac{(2\nu+1)\pi}{2n}}{\sin\left(\dfrac{1}{2m}-\dfrac{2\nu+1}{2mn}\right)\pi} + \frac{1}{n}, & \text{if } n = 2t+1.\end{cases}$$

We note here that for the projection obtained by averaging trigonometric polynomial interpolants, Morris and Cheney [14] developed the same norm as in (4.2) for the case where n is odd.

The proof of Theorem 1 depends on three interesting lemmas discussed below.

Lemma 4.1. Let f be defined on γ. Then for $z \in \gamma$, $z = e^{i\alpha}$, $-\frac{2\mu\pi}{mn} \leq \alpha \leq \frac{2(m-\mu)\pi}{mn}$, $0 \leq \mu \leq m-1$, we have

$$\left|L_{n-1}^{(\phi_\mu)}(z; f)\right| \leq \|f\| \, z^{-(n-1)/2} \, e^{-\mu\pi i/m} L_{n-1}^{(\phi_\mu)}(z; g), \tag{4.3}$$

where $g(z) := z^{-1/2}$.

Note that $L_{n-1}^{(\phi_\mu)}(z; g)$ behaves somewhat like the absolute value function since the right-hand side of (4.3) is always real. Indeed this expression is real for all $z \in \gamma$. We only provide a proof for the case $\mu = 0$. A complete proof of Lemma 4.1 can be found in [15].

Proof of Lemma 4.1. Let $\theta_\nu := \arg(\omega^\nu)$ and recall that

$$L_{n-1}(z; f) = \frac{z^n - 1}{n} \sum_{\nu=1}^{n} \frac{\omega^\nu f(\omega^\nu)}{z - \omega^\nu}. \tag{4.4}$$

Hence

$$|L_{n-1}(z; f)| \leq \frac{1}{n} \|f\| \sum_{\nu=1}^{n} \left| \frac{z^n - 1}{z - \omega^\nu} \right|. \tag{4.5}$$

Since $|w|^2 = w\bar{w}$, we have

$$\left| \frac{z^n - 1}{z - \omega^\nu} \right|^2 = z^{-n+1} \omega^\nu \left(\frac{z^n - 1}{z - \omega^\nu} \right)^2. \tag{4.6}$$

A simple calculation shows that

$$\left(\frac{z^n - 1}{z - w^\nu} \right)^2 = \left(\frac{e^{in\phi/2} \sin(n\phi/2)}{e^{i(\phi+\theta_\nu)/2} \sin \frac{1}{2}(\phi - \theta_\nu)} \right)^2, \qquad \nu = 1, \ldots, n. \tag{4.7}$$

Since $\mu = 0$, we have by assumption that $0 \leq \phi \leq \frac{2\pi}{n}$. For such ϕ it is easy to see that $\frac{\sin(n\phi/2)}{\sin \frac{1}{2}(\phi - \theta_\nu)}$ is never positive for $\nu = 1, \ldots, n$. In view of this and since $(z^{-n})^{1/2} = -e^{-in\phi/2}$, we have from (4.6) and (4.7)

$$\left| \frac{z^n - 1}{z - \omega^\nu} \right| = -z^{-(n-1)/2} \omega^{\nu/2} \frac{z^n - 1}{z - \omega^\nu}. \tag{4.8}$$

Also

$$\left. \frac{z^n - 1}{z - \omega^\nu} \right|_{z=\omega^k} = \begin{cases} 0, & \text{if } k \neq \nu \\ n\omega^{\nu(n-1)}, & \text{if } k = \nu, \end{cases}$$

so

$$\frac{z^n - 1}{z - \omega^\nu} = n\omega^{-\nu} \ell_\nu(z), \tag{4.9}$$

where $\ell_\nu(z)$, $\nu = 1,\dots,n$, are the Lagrange polynomials associated with the n-th roots of unity:

$$\ell_\nu(z) := \prod_{\substack{k=1\\k\neq\nu}}^{n} \frac{z-\omega^k}{\omega^\nu-\omega^k}.$$

Using (4.9) in (4.8) yields

$$\left|\frac{z^n-1}{z-\omega^\nu}\right| = nz^{-(n-1)/2}\ell_\nu(z)\omega^{-\nu/2}. \tag{4.10}$$

In light of (4.5), the result (4.4) for the special case $\mu = 0$ now follows by summing (4.10) over μ and then dividing by n. ∎

The next two lemmas are stated without proof. The proofs can be found in [15].

Lemma 4.2. For $z \in \gamma$ and g as in Lemma 4.1, the following inequality holds

$$\left|L_{n-1}^{(\phi_\mu)}(z;g)\right| \le \frac{1}{n}\sum_{\nu=0}^{n-1} \csc\frac{(2\nu+1)\pi}{2n}. \tag{4.13}$$

This bound is realized only when $z = \epsilon_\mu := \exp\left[\left(\frac{m-2\mu}{mn}\right)\pi i\right]$, the point on the unit circle "midway" between ϵ^n and ϵ.

Note that Gronwall's theorem [11] follows immediately from Lemmas 4.1 and 4.2.

Lemma 4.3. For $z \in \gamma$ and $g(z) = z^{-1/2}$, the following inequality is valid:

$$\left|\frac{1}{m}\sum_{\mu=0}^{m-1} e^{-\mu\pi i/m} L_{n-1}^{(\phi_\mu)}(z;g)\right| \le \tau_{n-1}^{(m)}, \tag{4.14}$$

where $\tau_{n-1}^{(m)}$ is given by (4.2). This bound is realized only when $z = \epsilon := \exp(\pi i/mn)$.

Note that the left-hand side of (4.14) represents the average of that part of the bound in (4.3) which depends on μ. Moreover, if $m = 1$, then (4.14) reduces to a special case of Lemma 4.2.

For the proof of Theorem 1, we note first that in view of (2.5), (4.3) and (4.14), $\|M_{n-1,m}\|$ is clearly bounded by $\tau_{n-1}^{(m)}$. To see that this is the best possible bound, it is necessary to exhibit a function f with $\|f\| = 1$ for which the bound is attained. The construction of such a function, which in fact belongs to $A(\gamma)$, can be found in [15, §4].

We end this section with a remark about the relationship between $\tau_{n-1}^{(m)}$ and a generalized Lebesgue function. If we define $\Lambda_{n,m}(z) := \frac{1}{m}\sum_{\nu=1}^{mn}\left|\tilde{\ell}_\nu(z)\right|$, where $\tilde{\ell}_\nu(z)$, $\nu = 1,\dots,mn$, are the fundamental Lagrange polynomials associated with the (mn)-th roots of unity, and define $V_{n-1,m}(z)$ to be the quantity inside the absolute value lines in (4.14), then it follows ([15]) that $\Lambda_{n,m}(z) = z^{-(n-1)/2}V_{n-1,m}(z)$ for $z = e^{i\phi}$, $0 \le \phi \le \frac{2\pi}{mn}$. The function $\Lambda_{n,m}(z)$ is periodic with period $2\pi/mn$ so the above relationship has an obvious extension to all of γ. The function $\Lambda_{n,m}(z)$ can be thought of as a generalized Lebesgue function for the mean projection given by (2.5). Furthermore (see [15]), for $z \in \gamma$,

$$\lim_{m\to\infty} \Lambda_{n,m}(z) = \tau_{n-1}.$$

5. Asymptotic Results

In this section we study the behavior of $\tau_{n-1}^{(m)}$ as $m \to \infty$ and as $n \to \infty$. Note that, for n odd or even,

$$\tau_{n-1}^{(m)} = \frac{2}{mn}\sum_{\nu=0}^{t-1} \frac{\cot\frac{(2\nu+1)\pi}{2n}}{\sin\left(\frac{1}{2m} - \frac{2\nu+1}{2mn}\right)\pi} + o(1). \tag{5.1}$$

By definition of the definite integral

$$\lim_{n\to\infty} \frac{\pi}{n}\sum_{\nu=0}^{t-1}\left[\frac{\cot\frac{(2\nu+1)\pi}{2n}}{\sin\left(\frac{1}{2m} - \frac{2\nu+1}{2mn}\right)\pi} - \frac{\csc\frac{\pi}{2m}}{\frac{(2\nu+1)\pi}{2n}}\right] \tag{5.2}$$

$$= \int_0^{\pi/2}\left[\frac{\cot x}{\sin\left(\frac{\pi}{2m} - \frac{x}{m}\right)} - \frac{\csc(\pi/2m)}{x}\right]dx =: J_m.$$

A tedious calculation (see [15]) verifies that

$$J_m = \sum_{j=1}^{2m-1} \csc\frac{(2j-1)\pi}{2m} \ln\frac{\sin\left(\frac{\pi j}{2m}\right)}{\sin\left(\frac{2j-1}{4m}\right)\pi} + \csc\frac{\pi}{2m}\left(\ln\left(\frac{4m}{\pi}\sin\frac{\pi}{4m}\right)\right). \tag{5.3}$$

Then $\tau_{n-1}^{(m)}$ can be written as

$$\tau_{n-1}^{(m)} = \frac{2}{m\pi}\left(J_m + 2\csc\left(\frac{\pi}{2m}\right)\sum_{\nu=0}^{t-1}\frac{1}{2\nu+1}\right) + o(1). \tag{5.4}$$

Using the formula

$$\sum_{\nu=0}^{r}\frac{1}{2\nu+1} = \frac{1}{2}\ln r + \ln 2 + \frac{1}{2}\beta + o(1),$$

where β is Euler's constant, (5.4) becomes

$$\tau_{n-1}^{(m)} = \frac{2}{m\pi}\left(J_m + \csc\frac{\pi}{2m}(\ln n + \ln 2 + \beta)\right) + o(1). \tag{5.5}$$

In Table 1 we have listed the values of $\tau_{n-1}^{(m)}$ and τ_{n-1} for various choices of m and n. Consideration of these values suggests that $\lim_{m\to\infty} \tau_{n-1}^{(m)} = \tau_{n-1}$. Indeed as noted in [15], $\tau_{n-1}^{(m)}$ decreases monotonically to τ_{n-1} as $m \to \infty$, and the convergence has order $O(m^{-2})$.

TABLE 1

Values of $\tau_{n-1}^{(m)}$ and τ_{n-1}

m:	1	5	10	15	20	25	30	35	50	∞
$n-1$	$\tau_{n-1}^{(m)}$									τ_{n-1}
1	1.414	1.278	1.275	1.274	1.274	1.273	1.273	1.273	1.273	1.273
2	1.667	1.444	1.438	1.437	1.436	1.436	1.436	1.436	1.436	1.436
3	1.848	1.562	1.555	1.553	1.553	1.552	1.552	1.552	1.552	1.552
4	1.989	1.654	1.645	1.643	1.643	1.643	1.643	1.642	1.642	1.642
5	2.104	1.729	1.719	1.717	1.717	1.716	1.716	1.716	1.716	1.716
6	2.202	1.792	1.782	1.780	1.779	1.779	1.779	1.779	1.778	1.778
7	2.287	1.847	1.836	1.834	1.833	1.833	1.833	1.833	1.833	1.832
8	2.362	1.896	1.884	1.882	1.881	1.881	1.881	1.880	1.880	1.880
9	2.429	1.939	1.927	1.925	1.924	1.923	1.923	1.923	1.923	1.923
10	2.489	1.978	1.966	1.963	1.962	1.962	1.962	1.962	1.962	1.961
20	2.901	2.245	2.229	2.226	2.225	2.224	2.224	2.224	2.224	2.223
30	3.149	2.405	2.387	2.384	2.383	2.382	2.382	2.382	2.381	2.381
50	3.466	2.610	2.590	2.586	2.585	2.584	2.584	2.583	2.583	2.583
100	3.901	2.892	2.868	2.863	2.862	2.861	2.861	2.860	2.860	2.860
200	4.339	3.175	3.148	3.143	3.141	3.140	3.140	3.139	3.139	3.139
500	4.920	3.551	3.519	3.513	3.511	3.510	3.510	3.510	3.509	3.509
1000	5.360	3.836	3.801	3.794	3.792	3.791	3.790	3.790	3.789	3.789

It follows from the above discussion and (5.5) that

$$\lim_{m\to\infty} \tau_{n-1}^{(m)} = \tau_{n-1} = \frac{4}{\pi^2}\ln n + \eta + o(1),$$

where $\eta = 0.98941\ldots$ and $o(1)$ tends to zero monotonically as n increases indefinitely. (See also [10].)

Using $m = 1$ in (5.5) yields

$$\tau_{n-1}^{(1)} = \frac{2}{\pi}\ln n + \frac{2}{\pi}\left(\beta + \ln\frac{8}{\pi}\right) + o(1)$$

which is Gronwall's asymptotic formula [11].

6. Approximating Laurent Series

For $1 \le p \le \infty$, let $X^{(p)}$ be the Banach space of complex-valued functions which are L_p-integrable on γ with respect to $|dz| :=$ unit of arc length. Let r and s be nonnegative integers such that $r + s + 1 = n$, where, as before, n denotes a positive integer. Let

(6.1) $$c_k := \langle f, z^k \rangle, \qquad k \text{ an integer},$$

where $\langle f, g \rangle := \int_\gamma f(z) g(z)\, |dz|$, and define the Laurent projection $Q_{r,s}$ by

(6.2) $$(Q_{r,s} f)(z) := \sum_{k=-r}^{s} c_k\, z^k .$$

Note that if f has a Laurent series expansion, then $Q_{r,s}$ represents a particular section of this expansion of degree s in z and degree r in z^{-1}. Several interesting results along these lines can be found in [4,13].

In this section we will use the mean projections to approximate (6.2). For the purpose of motivation, suppose that f is Riemann integrable on γ. Recall the proof of Proposition 3.1 and consider

(6.3) $$A_{n-1,m}(z; f_r) = \frac{1}{mn} \sum_{j=0}^{mn-1} \sum_{k=0}^{n-1} \varsigma^{-j(k-r)} f\left(\varsigma^j\right) z^k,$$

where $f_r(z) := z^r f(z)$. If we form the product $z^{-r} A_{n-1,m}(z; f_r)$, we note that the coefficient for z^μ, $-r \le \mu \le s$, in this product is

(6.4) $$\hat{c}_\mu := \frac{1}{mn} \sum_{j=0}^{mn-1} \varsigma^{-j\mu} f\left(\varsigma^j\right).$$

This last sum is easily seen to be the mn-point trapezoidal approximation to (6.1). Hence $z^{-r} A_{n-1,m}(z; f_r)$ can be viewed as an approximation to (6.2).

Next, let X be the space of all functions f which have an absolutely convergent Laurent expansion $f(z) = \sum_{k=-\infty}^{\infty} c_k\, z^k$. Then

(6.5) $$z^r f_r(z) = \sum_{\nu=0}^{\infty} \sum_{\mu=0}^{mn-1} c_{\mu+\nu mn-r}\, z^{\mu+\nu mn} + \sum_{\nu=1}^{\infty} \sum_{\mu=0}^{mn-1} c_{\mu-\nu mn-r}\, z^{\mu-\nu mn}.$$

Since $M_{n-1,m} = T_{n-1} F_{mn-1}$ and $F_{mn-1}\left(z; z^{\mu\pm\nu mn}\right) = z^\mu$, we have a projection $\tilde{Q}_{r,s,m}$ defined by

(6.6) $$\left(\tilde{Q}_{r,s,m} f\right)(z) := z^{-r} A_{n-1,m}(z; f_r)$$

$$= \sum_{\nu=0}^{\infty} \sum_{\mu=-r}^{s} c_{\mu+\nu mn}\, z^{\mu} + \sum_{\nu=1}^{\infty} \sum_{\mu=-r}^{s} c_{\mu-\nu mn}\, z^{\mu}.$$

For such f it follows that

$$\hat{c}_\mu = c_\mu + \sum_{\nu=1}^{\infty} (c_{\mu+\nu mn} + c_{\mu-\nu mn}), \qquad -r \le \mu \le s. \tag{6.7}$$

Two properties of the projection $\tilde{Q}_{r,s,m}$ follow immediately.

Proposition 6.1. Let $f \in X^{(2)} \cap X$. Then $\tilde{Q}_{r,s,m} f$ converges to $Q_{r,s} f$ in the mean as $m \to \infty$.

Proof. Let $f \in X^{(2)} \cap X$ and consider

$$\left\| \left(Q_{r,s} - \tilde{Q}_{r,s,m}\right) f \right\|_2 = \left[\sum_{\mu=0}^{n-1} \left| \sum_{\nu=1}^{\infty} c_{\mu+\nu mn} + c_{\mu-\nu mn} \right|^2 \right]^{1/2}$$

$$\le \left[\sum_{\mu=0}^{n-1} \left| \sum_{\nu=1}^{\infty} c_{\mu+\nu mn} \right|^2 + \sum_{\mu=0}^{n-1} \left| \sum_{\nu=1}^{\infty} c_{\mu-\nu mn} \right|^2 \right]^{1/2}$$

$$\le \left[\sum_{\mu=0}^{n-1} \sum_{\nu=1}^{\infty} |c_{\mu+\nu mn}|^2 + \sum_{\mu=0}^{n-1} \sum_{\nu=1}^{\infty} |c_{\mu-\nu mn}|^2 \right]^{1/2}.$$

Since $\|f\|_2 = \left(\sum_{j=-\infty}^{\infty} |c_j|^2 \right)^{1/2}$, the last expression above can be bounded by $c(m)\, \|f\|_2$ where $c(m) \to 0$ as $m \to \infty$. ■

We conclude this paper with the following proposition. We omit the proof.

Proposition 6.2. If $f \in X$, then $\tilde{Q}_{r,s,m} f \to Q_{r,s} f$ as $m \to \infty$. Moreover, for the class of functions analytic in an annulus containing γ, $\left\| Q_{r,s} - \tilde{Q}_{r,s,m} \right\| \to 0$ as $m \to \infty$.

REFERENCES

1. Brutman, L. (1982): *A sharp estimate of the Landau constants.* J. Approx. Theory **34**, 217-220.

2. Brutman, L. (1980): *On the polynomials and rational projections in the complex plane.* SIAM J. Numer. Anal. **17**, 366-372.

3. Brutman, L., and Pinkus, A. (1980): *On the Erdös conjecture concerning minimal norm interpolation on the unit circle.* SIAM J. Numer. Anal. **17**, 373-375.
4. Chalmers, B.L. and Mason, J.C. (1984): *Minimal L_p projections by Fourier, Taylor and Laurent Series.* J. Approx. Theor. **40**, 293-297.
5. Dienes, P. (1957): The Taylor Series, Dover Publications, New York.
6. Erdös, P. (1968): *Problems and results on the convergence and divergence of the Lagrange interpolation polynomials and some external problems.* Mathematics (Cluj) **10**, 64-73.
7. Fejer, L. (1910): *Lebesguesche Konstanten und divergente Fourierreihe.* J. für die Reine und angew. Math. **138**, 22-53.
8. Galkin, P.V. (1971): *Estimates for the Lebesgue constants.* Proc. Steklov Inst. Math. **109**, 1-3.
9. Geddes, K.O. (1978): *Near-minimax polynomial approximation in an elliptical region.* SIAM J. Numer. Anal. **15**, 1225-1233.
10. Geddes, K.O. and Mason, J.C. (1975): *Polynomial approximation by projections on the unit circle.* SIAM J. Numer. Anal. **12**, 111-120.
11. Gronwall, T.H. (1921): *A sequence of polynomials connected with the n-th roots of unity.* Bull. Amer. Math. Soc. **27**, 275-279.
12. Mason, J.C. (1981): *Near-minimax interpolation by a polynomial in z and z^{-1} on a circular annulus.* IMA J. of Numer. Anal. **1**, 359-367.
13. Mason, J.C. and Chalmers, B.L. (1984): *Near-best L_p approximations by Fourier, Taylor, and Laurent series.* IMA J. of Numer. Anal. **4**, 1-8.
14. Morris, P.D. and Cheney, E.W. (1973): *Stability properties of trigonometric interpolating operators.* Math. Z. **131**, 153-164.
15. Palagallo, J.A. and Price, T.E. (1987): *Near-best approximation by averaging polynomial interpolants.* IMA J. of Numer. Anal. **7**, 107-122.
16. Rivlin, T.J. (1982): *On Walsh overconvergence.* J. Approx. Theor. **36**, 334-345.
17. Watson, G.N. (1930): *The constants of Landau and Lebesgue.* Q.J. Math. Oxford Ser. **1**, 310-318.

Boundary Collocation in Fejér Points for Computing Eigenvalues and Eigenfunctions of the Laplacian

Lothar Reichel
University of Kentucky
Department of Mathematics
Lexington, KY 40506

Abstract. The boundary collocation method is applied to the computation of eigenvalues and eigenfunctions of the Laplace operator on planar simply connected regions with smooth boundaries. For convex regions we seek to approximate the eigenfunctions by a linear combination of basis functions that contain Bessel functions of the first kind. Our method differs from related schemes proposed previously in that we distribute the collocation points differently, and we use a different iterative scheme for computing eigenvalues and eigenfunctions. This makes our method both faster and more accurate. For nonconvex regions rapid convergence generally can be achieved only if the eigenfunctions are approximated by functions with singular points in the finite plane. A boundary collocation method with such basis functions is also described.

1. Introduction

Let Ω be a simply connected bounded open set in the complex plane $\mathbb{C}$ with smooth boundary $\partial\Omega$. We describe an application of the *boundary collocation method* (BCM) to the computation of eigenfunctions u and eigenvalues λ of the Laplacian on Ω,

(1.1a) $\quad \Delta u + \lambda u = 0 \quad \text{in } \Omega,$

(1.1b) $\quad u = 0 \quad \text{on } \partial\Omega,$

(1.1c) $\quad \|u\|_2 := \iint\limits_{\Omega} u^2 d\Omega = 1\,.$

Twenty years ago the BCM was employed successfully by Fox, Henrici and Moler [FHM] to compute accurate approximate solutions of (1.1) on ellipses and an L-shaped region. Since then the solution of (1.1) by boundary collocation or related methods has continued to attract attention, see, e.g., [Mo], [BN], [Fi], [HZ]. A recent survey of numerical methods for (1.1), among them the BCM, is given by Kuttler and Sigillito [KS].

In the engineering literature boundary collocation is also known as *point matching,* and Bates and Ng [BN, p. 155] state "The point matching or collocation method is an attractively economic technique, from the point of view of programming effort and computer time, for calculating the conditions for resonance in continuous physical systems, provided the vibrations can be described by partial differential equations." Moreover, the BCM allows the determination of error bounds for the computed approximate eigenvalues and eigenfunctions [FHM], [MP], [Mo], [Fi].

Our implementation of the BCM differs from those in [FHM], [Mo], [BN], [Fi] in that we use a faster iterative scheme to compute the eigenvalues and eigenfunctions. This is described in Section 2. Moreover, we choose Fejér points, defined below, as collocation points. This choice yields higher accuracy than previously obtained, and, in fact, gives errors close to the optimal errors achieved by techniques based on semi-infinite programming [HZ]. The latter techniques require more computational work than boundary collocation. Numerical examples are presented in Section 3. The subspace of functions used to approximate eigenfunctions in [FHM], [Mo], [BN], [Fi], [HZ], as well as in Sections 2-3, yields poor accuracy for pronouncedly nonconvex Ω [Re1]. In Section 4, we choose a subspace of functions suitable for such domains and present computed examples. This subspace does, however, require substantially more computational work than the subspace of Sections 2-3. The numerical scheme of Section 4 may nevertheless compare favorably to schemes requiring evaluation of integrals with singular kernels, such as the scheme described in [SH].

We next describe the BCM in some detail and introduce notation to be used later. Following [FHM], [Mo], [Fi] we approximate the eigenfunctions by a linear combination

$$u_{2l-1}(z) := a_o J_o(\lambda^{1/2} r) + \sum_{k=1}^{l-1} J_k(\lambda^{1/2} r)\big(a_k \cos(k\theta) + b_k \sin(k\theta)\big), \tag{1.2}$$

where $z = x+iy$, $i=\sqrt{-1}$, and J_k denotes the Bessel function of the first kind of order k. u_{2l-1} can be evaluated rapidly by using recurrence relations for the J_k. For any λ and coefficients a_k, $b_k \in \mathbb{R}$, $u_{2l-1}(x,y) := u_{2l-1}(z)$ satisfies (1.1a) but generally not (1.1b,c). In the BCM one obtains an approximate solution of (1.1) by selecting *collocation points* $z_j = r_j e^{i\theta_j} \in \partial\Omega$, $1 \le j \le m$, where $m \ge n := 2l-1$, and determining λ and the a_k, b_k so that $\sum_{j=1}^{m} (u_{2l-1}(z_j))^2$ is small and $\|u_{2l-1}\|_2 \ge 1$. These conditions can be satisfied only if the columns of the $m \times n$ *collocation matrix* $A = A(\lambda) = [\alpha_{jk}(\lambda)]$,

$$\begin{cases} \alpha_{j,2k} := J_k(\lambda^{1/2} r_j)\cos(k\theta_j), & 0 \le k < l, \quad 1 \le j \le m, \\ \alpha_{j,2k+1} := J_k(\lambda^{1/2} r_j)\sin(k\theta_j), & 1 \le k < l, \quad 1 \le j \le m, \end{cases} \tag{1.3}$$

are nearly linearly dependent. In the computations, we scale the J_k so that the columns of A have Euclidean length one.

Let $\mu(\lambda)$ be the smallest eigenvalue of $B(\lambda) := A^T(\lambda)A(\lambda)$. We first determine rough estimates of local minima of $\mu(\lambda)$ by determining the singular value decomposition of $A(\lambda)$ for some equidistant values of λ. Accurate approximations of local minima of $\mu(\lambda)$ are then determined by a scheme related to inverse iteration, where we, however, only need to determine the LU-decomposition of $B(\lambda)$ for few values of λ. Let $\mu(\lambda^*)$ be a local minimum of $\mu(\lambda)$. If $\mu(\lambda^*)$ is small then the columns of A are nearly linearly dependent, and λ^* is an approximate eigenvalue of (1.1). By methods described in [MP], [Mo], one can determine how close λ^* is to an eigenvalue of (1.1), see Section 3.

Let $\psi(\omega)$ be a conformal mapping from $\{\omega: |\omega|>1\}$ to $\mathbb{C}\backslash(\Omega\cup\partial\Omega)$ such that $\psi(\infty)=\infty$. ψ can be continued continuously to a bijective mapping from $\{\omega: |\omega|\geq 1\}$ to $\mathbb{C}\backslash\Omega$. We denote this extension also by ψ. The points

$$z_j := \psi(e^{2\pi ij/m}),\ \ 0\leq j<m,$$

are called *Fejér points* [Cu], [Ga]. The appropriateness of collocation at Fejér points for the present problem was discussed in [Re1]. Collocation at Fejér points has previously been suggested by Curtiss [Cu] for the solution of Dirichlet problems for the Laplace equation with harmonic polynomials as basis functions. m Fejér points can be computed in $O(m \log m)$ arithmetic operations by methods described in [Gu]. Recent reviews of numerical methods for conformal mapping are found in [He], [T]. For our purposes the Fejér points need not be determined very accurately.

2. Iterative Computation of Eigenvalues and Eigenfunctions

We wish to determine local minima of $\mu(\lambda)$, the smallest eigenvalue of the symmetric positive semi-definite $n\times n$ matrix $B=B(\lambda)$. Values λ^* for which $\mu(\lambda)$ achieves a sufficiently small local minimum are approximate eigenvalues of (1.1). By tabulating $\mu(\lambda)$ we can determine a triple $\{\lambda_1^0,\lambda_2^0,\lambda_3^0\}$, $\lambda_1^0<\lambda_2^0<\lambda_3^0$, such that $\mu(\lambda)$ has precisely one local minimum λ^* in the open interval $(\lambda_1^0,\lambda_3^0)$ and $\mu(\lambda_2^0)<\min\{\mu(\lambda_1^0),\mu(\lambda_3^0)\}$. We then interpolate $\mu(\lambda)$ by a quadratic $q(\lambda)$ at λ_j^0, $1\leq j\leq 3$, and determine λ_4^0, the zero of $q'(\lambda)$. From $\{(\lambda_j^0,\mu(\lambda_j^0))\}_{j=1}^4$ we can determine a new triple $\{\lambda_1^1,\lambda_2^1,\lambda_3^1\}$, $\lambda_1^1<\lambda_2^1<\lambda_3^1$, such that $\lambda_3^1-\lambda_1^1<\lambda_3^0-\lambda_1^0$ and $\lambda^*\in(\lambda_1^1,\lambda_3^1)$. These iterations are repeated until

$$\lambda_3^j-\lambda_1^j<\tilde{\epsilon}\,, \tag{2.1}$$

for some given $\tilde{\epsilon}>0$, or until we switch to a different iteration scheme, see below.

We determine $\mu(\lambda)$ by computing the singular value decomposition of $A(\lambda)$. Note that $\mu(\lambda)$ is an analytic function of λ, while the smallest singular value of $A(\lambda)$ is not. The above iteration scheme is related to, but appears to be different from, a scheme proposed by Moler [Mo] for computing accurate approximations of local minima of $\mu(\lambda)$.

The elements of $A(\lambda)$ can be evaluated rapidly using a three-term recurrence relation for the J_k. The time-consuming part of the above iterative scheme is the repeated singular value decomposition of $A(\lambda)$ for many close values of λ. When $\lambda_3^j-\lambda_1^j$ is small, we therefore switch to an iterative scheme related to inverse iteration. This scheme is less well-conditioned than the computation of the smallest singular value of $A(\lambda)$, and we therefore also switch to higher precision arithmetic.

For some $j\geq 0$, to be specified below, define $B_0 := B(\lambda_2^j)$, $\delta B = \delta B(\lambda) := B(\lambda)-B_0$ and $\mu_0 := \mu(\lambda_2^j)$. Let x_0 be an eigenvector of B_0 belonging to μ_0 such that $x_0^Tx_0=1$. x_0 can be determined from the singular value decomposition of $A(\lambda_2^j)$. We seek $\delta\mu_0\in\mathbb{R}$, $\delta x_0\in\mathbb{R}^n$ such that

$$\begin{cases} B(x_0+\delta x_0) = (x_0+\delta x_0)(\mu_0-\delta\mu_0) \\ (x_0+\delta x_0)^T(x_0+\delta x_0) = 1\,, \end{cases}$$

or equivalently, with $r := Bx_0-\mu_0 x_0$,

$$(2.2)\qquad \begin{cases} (B_0-\mu_0 I)\delta x_0+x_0\delta\mu_0 = -\delta B\,\delta x_0-\delta x_0\delta\mu_0-r \\ x_0^T\delta x_0 = -\frac{1}{2}\delta x_0^T\delta x_0\,. \end{cases}$$

Equation (2.2) suggests the iterative scheme

$$(2.3)\qquad \begin{pmatrix} B_0-\ \mu_0 I & x_0 \\ x_0^T & 0 \end{pmatrix}\begin{pmatrix} \delta x_0^{(k+1)} \\ \delta\mu_0^{(k+1)} \end{pmatrix} = \begin{pmatrix} -\delta B\,\delta x_0^{(k)} - \delta x_0^{(k)}\delta\mu_0^{(k)}-r \\ -\frac{1}{2}\delta x_0^{(k)T}\ \delta x_0^{(k)} \end{pmatrix}$$

Let $\mu_0<\mu_1\leq\mu_2\leq\cdots\leq\mu_{n-1}$ be the eigenvalues of B_0 and $x_0,x_1,\ldots,x_{n-1}$ the corresponding eigenvectors. Then

$$\tilde{B} := \tilde{B}(\lambda) := \begin{pmatrix} B_0-\mu_0 I & x_0 \\ x_0^T & 0 \end{pmatrix}$$

has the eigenvalues 1,-1 and $\mu_j-\mu_0$, $1\leq j<n$, and the corresponding eigenvectors $(x_0,1)^T$, $(x_0,-1)^T$ and $(x_j,0)^T$, $1\leq j<n$. In particular, $\tilde{B}$ is nonsingular. The iterative scheme (2.3) converges if $\delta B,\delta x_k$ and $\delta\mu_k$ are small compared to $\mu_1-\mu_0$. Note that $\mu_0=\mu_0(\lambda_2^j)$, $\mu_1=\mu_1(\lambda_2^j)$ are known from the singular value decomposition of $A(\lambda_2^j)$. Moreover, let $\delta A := A(\lambda)-A(\lambda_2^j)$. Then $\|\delta B\|_F\lesssim 2\|A(\lambda_2^j)\|_F\|\delta A\|_F$, where $\|\ \|_F$ denotes the Frobenius norm. By the column scaling of A, we have $\|A\|_F=n^{1/2}$, and we estimate $\|\delta A\|_F$ by $\|A(\lambda_3^j)-A(\lambda_1^j)\|_F$. In the numerical examples of Section 3 we switched to iterations (2.3) when

$$(2.4)\qquad 2n^{1/2}\|A(\lambda_3^j)-A(\lambda_1^j)\|_F\leq(\mu_1-\mu_0)/10\,.$$

Iterations (2.3) yield $(\mu_0-\delta\mu_0)(\lambda)$ and $(x_0+\delta x_0)(\lambda)$. We determine λ such that $(\mu_0-\delta\mu_0)(\lambda)=0$ by the secant method. The so computed values of λ are approximate eigenvalues of (1.1) and error bounds can be determined.

If $\mu_0\approx\mu_1$, then (2.3) will not converge. Formulas analogous to (2.3) can be derived for multiple or close eigenvalues, but these formulas and the resulting computer program become more complicated, and we feel that the use of these analogues of (2.3) is not worthwhile. We note that the smallest eigenvalue of the Laplacian is always simple by the nodal line theorem [CH].

Previously described methods for computing local minima of $\mu(\lambda)$, see [FHM], [BN], [Mo], [Fi], determine an LU-decomposition or a QR or SVD factorization of $A(\lambda)$ for each λ.

3. Numerical Experiments

This section contains numerical experiments with the expansion (1.2). Let $\lambda^* \geq 0$ be a computed approximate eigenvalue and let u_{2l-1}^* be the corresponding computed eigenfunction. Moler and Payne [MP] show how error bounds for λ^* and u_{2l-1}^* can be computed. We just state their result on bounds for λ^*:

Lemma 3.1 ([MP]). Let

$$\epsilon := \max_{z \in \partial\Omega} |u_{2l-1}^*(z)| \cdot (\text{area } \Omega)^{1/2} \cdot \left(\iint_{\Omega} (u_{2l-1}^*)^2 d\Omega \right)^{-1}, \tag{3.1}$$

and assume $\epsilon < 1$. Then there is an eigenvalue λ of (1.1) in the interval $\dfrac{\lambda^*}{1+\epsilon} \leq \lambda \leq \dfrac{\lambda^*}{1-\epsilon}$. □

Following [FHM], [Mo], [Fi] we assume in the present section that $0 \in \Omega$ and let D_0 be the largest disk inscribed in Ω with center 0. (3.1) can be replaced by

$$\epsilon := \max_{z \in \partial\Omega} |u_{2l-1}^*(z)| (\text{area } \Omega)^{1/2} \left(\iint_{D_0} (u_{2l-1}^*)^2 d\Omega \right)^{-1}, \tag{3.1'}$$

where the double integral can be evaluated analytically [FHM],[Mo],[Fi]. Let $z(t) = x(t) + iy(t)$, $0 \leq t < 2\pi$, be a parametric representation of $\partial\Omega$. We evaluate

$$\text{area } \Omega = \int_0^{2\pi} x(t) y'(t) dt$$

by the trapezoidal rule.

The computations were carried out on a VAX 11/750 in double precision arithmetic (17 significant digits), except iterations (2.3) which were carried out in quadruple precision (34 significant digits). $\tilde{\epsilon}$ in (2.1) was chosen to be $1 \cdot 10^{-15}$. In all examples (2.4) was satisfied before (2.1). The secant iterations were terminated when the change in λ was of magnitude $\leq 1 \cdot 10^{-15}$.

Example 3.1. Let Ω be an ellipse with boundary curve $z(t) = a\cos(t) + i\sin(t)$, $0 \leq t < 2\pi$. We wish to compute the lowest eigenvalue, and, due to symmetry with respect to the x- and y-axes of the corresponding eigenfunction, we seek $u_n(z)$ of the form

$$u_n(z) = \sum_{k=0}^{n-1} a_k J_{2k}(\lambda^{1/2} r) \cos(2k\theta), \quad z = re^{i\theta}, \tag{3.2}$$

and allocate collocation points in the first quadrant only. m Fejér points in the first quadrant are given by

$$z_j = z\big((2j-1)\pi/(4m)\big), \quad 1 \leq j \leq m. \tag{3.3}$$

We compare results obtained for Fejér points with results previously obtained for nodes equidistant with respect to arc length in [FHM], [Fi]. In the tables the latter are called equally spaced. Upper and lower bound refer to the bounds for the eigenvalue obtained by Lemma 3.1 with ϵ given by (3.1′).

a	n	m	nodes	ϵ	upper bound lower bound
2	5	5	Fejér	$4.5 \cdot 10^{-10}$	3.56672660_{13}^{46}
2	5	5	equally spaced	$1.8 \cdot 10^{-9}$	3.566726_{59}^{61}
2	5	10	Fejér	$4.1 \cdot 10^{-10}$	3.56672660_{14}^{44}
2	5	10	equally spaced	$5.0 \cdot 10^{-10}$	3.56672660_{11}^{48}

Table 3.1

For $n=m$ the Fejér points yield an ϵ roughly a factor five times smaller than the value obtained with equally spaced nodes, and this ratio decreases if $m=2n$. This behavior was noticed in many other experiments, also. Our error for $n=m=5$ and equally spaced nodes is slightly less than the error achieved in [FHM, Table 1] with $n=m=10$ and equally spaced nodes, indicating that our method for iteratively computing eigenvalues is not only faster but also more accurate than the method used in [FHM]. □

Example 3.2. This example differs from Example 3.1 only in that $a=4$.

a	n	m	nodes	ϵ	upper bound lower bound
4	5	5	Fejér	$8.6 \cdot 10^{-6}$	2.9202_{25}^{77}
4	5	5	equally spaced	$4.2 \cdot 10^{-5}$	2.920_{12}^{38}
4	5	10	Fejér	$5.1 \cdot 10^{-6}$	2.9202_{34}^{68}
4	5	10	equally spaced	$9.9 \cdot 10^{-6}$	2.9202_{22}^{80}

Table 3.2

Again Fejér nodes yield better results, and choosing $m=2n$ improves the accuracy. If $n=5$, $m=20$ and Fejér nodes are used, then almost no improvement in accuracy was achieved when compared to the case $n=5$, $m=10$. □

Example 3.3. This example should be compared to [HZ, Table 1]. The boundary is the same ellipse as in Example 3.1. We use expansion (3.2) with $m=5$ and $m=6$ and 10 Fejér points (3.3) in the first quadrant. The eigenvalues in Table 3.3 are the first four in the sequence of eigenvalues with eigenfunctions symmetric with respect to both x- and y-axes.

n	m	ϵ	upper bound lower bound
5	10	$4.1 \cdot 10^{-10}$	3.56672660_{14}^{44}
5	10	$3.3 \cdot 10^{-5}$	10.028_{07}^{74}
5	10	$2.4 \cdot 10^{-2}$	$2_{0.3}^{1.4}$
5	10	$2.6 \cdot 10^{-3}$	$24._{82}^{96}$

Table 3.3

n	m	ϵ	upper bound lower bound
6	10	$1.2 \cdot 10^{-13}$	3.56672660292_{84}^{94}
6	10	$8.1 \cdot 10^{-8}$	10.02840_{08}^{25}
6	10	$3.1 \cdot 10^{-4}$	20.8_{41}^{56}
6	10	$3.3 \cdot 10^{-5}$	24.88_{43}^{66}

Table 3.4

Tables 3.3-4 indicate rapid convergence for increasing n. ϵ in Tables 3.3-4 corresponds to $\mu(\hat{\lambda})$ in [HZ, Table 1]. This shows that the error bounds achieved in Table 3.3 are nearly as good as error bounds obtained in [HZ] using expansion (3.2) with $n=5$ and a method based on semi-infinite programming techniques. (In fact, we obtain a better bound for the smallest eigenvalue.) Comparing Table 3.4 to [HZ, Table 1], we find that for $n=6$ we obtain error bounds which are smaller than those obtained by semi-infinite programming techniques with $n=5$. □

Example 3.4. We compute the lowest eigenvalue of ellipses with parameter a large and choose $n=10$ in (3.2) and 20 Fejér points (3.3) in the first quadrant.

a	ϵ	upper bound lower bound
10	$7.4 \cdot 10^{-10}$	2.63273186^{49}_{09}
12	$1.3 \cdot 10^{-7}$	2.603938^{76}_{08}
15	$7.8 \cdot 10^{-5}$	2.575^{75}_{59}
19	$4.6 \cdot 10^{-3}$	2.5^{57}_{47}
20	$6.1 \cdot 10^{-3}$	2.5^{55}_{41}

Table 3.5

The accuracy of Table 3.5 compares favorably with the accuracy of [FHM, Table 1] and [Fi, Table 4.4a]. □

Example 3.5. Let Ω be the bounded region with boundary $z(t)=e^{it}-\frac{1}{8}e^{-3it}$, $0 \leq t < 2\pi$. Ω has approximately the shape of a square with sides of length 7/4 and center 0. We determine the lowest eigenvalue using an expansion (3.2). m Fejér points in the first quadrant are given by (3.3).

n	m	nodes	ϵ	upper bounds lower bounds
10	10	Fejér	$3.8 \cdot 10^{-8}$	6.448609^{72}_{22}
10	10	equally spaced	$2.4 \cdot 10^{-7}$	6.4486^{12}_{07}
10	20	Fejér	$3.8 \cdot 10^{-8}$	6.448609^{72}_{22}
10	20	equally spaced	$5.0 \cdot 10^{-8}$	6.448609^{80}_{14}

Table 3.6

□

Example 3.6. Let Ω be the nonsymmetric region bounded by the curve $z(t)=\frac{3}{2}e^{it}+\frac{1}{2}e^{-it}+\frac{1+i}{20}(e^{-2it}+1)$, $0 \leq t < 2\pi$. We now use expansion (1.2). m Fejér points are given by $z_j=z(2\pi(j-1)/m)$, $1 \leq j \leq m$. Table 3.7 shows the lowest eigenvalue.

n	m	nodes	ϵ	upper bounds lower bounds
21	21	Fejér	$9.9 \cdot 10^{-7}$	3.5906_{06}^{15}
21	21	equally spaced	$2.8 \cdot 10^{-6}$	3.5906_{00}^{21}
21	42	Fejér	$4.4 \cdot 10^{-7}$	3.5906_{08}^{13}
21	42	equally spaced	$5.1 \cdot 10^{-7}$	3.5906_{08}^{13}

Table 3.7

□

Example 3.7. This example considers a region Ω with a non-analytic boundary curve. Let Ω be the sports ground-like region obtained by splitting the unit disk into halves and placing a 2×2 square between the halves. We construct a parametric representation of $\partial\Omega$ using a vertical projection from the smallest circle enclosing Ω onto the boundary $\partial\Omega$. The parametric representation of $\partial\Omega$ in the first quadrant is given by $z(t) := x(t) + iy(t)$, where

$$\begin{cases} x(t) := & 2\cos(t), \qquad\qquad\qquad\qquad 0 \le t < 2\pi \\ y(t) := & \begin{cases} \sin(\arccos(2\cos(t)-1)) & 0 \le t < \pi/3 \\ 1, & \frac{\pi}{3} \le t \le \pi/2 \,. \end{cases} \end{cases}$$

The remaining part of the boundary is defined by symmetry. For a figure, see [Re3, Fig. 19]. We use expansion (3.2) and the nodes (3.3). These nodes are not Fejér points, but the parametric representation $z(t)$ is such that the distribution function for the z_j is close to the distribution function for the Fejér points.

n	m	ϵ	upper bounds lower bounds
11	22	$7.2 \cdot 10^{-3}$	$3.{}_{15}^{21}$

Table 3.8

The accuracy is lower than for analytic boundary curves. □

We conclude this section with some remarks on the method described. In all computed examples the use of Fejér points as collocation points gave better accuracy than nodes equidistant with respect to arc length. Generally, the Fejér points are not explicitly known, but can be computed rapidly. They need not be known to high accuracy. When n or λ increase it becomes more difficult to find local minima of $\mu(\lambda)$ through tabulation. The step length in the tabulation may have to be fairly small to detect a minimum. A priori inequalities for eigenvalues have been used in [Fi] to roughly determine locations of eigenvalues. See also [KS].

We applied the BCM with expansion (1.2) and with Fejér points as collocation points to determine the lowest eigenvalue for the pronouncedly nonconvex region Ω of Example 4.1 below. It was difficult to detect a local minimum of $\mu(\lambda)$ and impossible to achieve $\epsilon<1$ in (3.1'). We therefore in Section 4 describe a BCM using approximating functions with singularities in the finite plane.

4. Basis Functions with Singular Points in the Finite Plane

In this section we present numerical experiments with a BCM for computing eigenvalues for nonconvex Ω. Let $C^{p,\gamma}(\Omega\cup\partial\Omega)$ denote the set of functions whose p^{th} derivative is Hölder continuous on $\Omega\cup\partial\Omega$ with Hölder constant γ. Let for some $p+\gamma>0$, $u\in C^2(\Omega)\cap C^{p,\gamma}(\Omega\cup\partial\Omega)$, and let u, moreover, satisfy (1.1a) for some constant λ. By Eisenstat [Ei, Theorem 4.2], u can for $z\in\Omega\cup\partial\Omega$ be represented by Vekua's integral operator V,

$$(4.1)\quad \begin{cases} u(z) = (Vq)(z), \\ (Vq)(z) := Re\left[q(z)-\dfrac{\lambda}{2}(\bar{z}-\bar{z}_0)\displaystyle\int_{z_0}^{z} q(\varsigma)\frac{J_1(\sqrt{\lambda(z-\varsigma)(\bar{z}-\bar{z}_0)})}{\sqrt{\lambda(z-\varsigma)(\bar{z}-\bar{z}_0)}}d\varsigma\right], \end{cases}$$

where q is a function analytic in Ω and $q\in C^{p,\gamma}(\Omega\cup\partial\Omega)$, z_0 is an arbitrary but fixed point in $\Omega\cup\partial\Omega$, and J_1 is the first order Bessel function of the first kind. The bars denote complex conjugation. q is uniquely determined by the requirement $Im(q(z_0))=0$.

Conversely, if q is analytic in Ω and $q\in C^{p,\gamma}(\Omega\cup\partial\Omega)$, then u defined by (4.1) lies in $C^2(\Omega)\cap C^{p,\gamma}(\Omega\cup\partial\Omega)$ and satisfies (1.1a) [Ei]. Hence, approximating solutions of (1.1) by functions that satisfy (1.1a) is equivalent to an approximation problem for analytic functions. In particular, if we let $z_0=0$, then V maps the set of polynomials of degree $\leq l$ with complex coefficients, and with imaginary part vanishing at z_0, bijectively onto the set spanned by the basis in (1.2) [Ve, §22].

In the collocation scheme of this section we obtain basis functions by letting q be rational functions with fixed poles in $\mathbb{C}\backslash(\Omega\cup\partial\Omega)$ and by mapping these rational functions onto solutions of (1.1a) by V. The collocation points and poles are distributed as follows. Let S be a simply connected open point set in $\mathbb{C}$ with a smooth boundary ∂S. Assume $(\Omega\cup\partial\Omega)\subseteq S$ and define $\hat{S}:=S\backslash(\Omega\cup\partial\Omega)$. Let $\hat{\psi}(\omega)$ be a conformal mapping which maps an annulus $\{\omega\colon 1<|\omega|<\delta\}$ onto S, where $\delta>1$ is uniquely defined. By the smoothness of ∂S

and $\partial\Omega$, we can continue $\hat{\psi}(\omega)$ continuously to a bijective mapping on $\{\omega : 1 \le |\omega| \le \delta\}$. This extension of $\hat{\psi}$ is also denoted by $\hat{\psi}$. We may assume that $\partial\Omega$ is the image of the unit circle under $\hat{\psi}$.

Define a set of *poles* $\{w_{jl}\}_{j=1}^{l-1} \subset \partial S$ by $w_{jl} := \hat{\psi}(\rho e^{2\pi i(j-1)/(l-1)})$. The w_{jl} determines the space Q_l of rational functions to be used,

$$Q_l := \operatorname{span}\{1,(z-w_{1l})^{-1},(z-w_{2l})^{-1}, \dots ,(z-w_{l-1,l})^{-1}\}. \tag{4.2}$$

Introduce a set of *auxiliary nodes* $\{\varsigma_{jl}\}_{j=1}^{l} \subset \partial\Omega$, defined by $\varsigma_{jl} := \hat{\psi}(e^{2\pi i(j-1)/l})$. We will use the following bases of Q_l and VQ_l, respectively,

$$\begin{cases} q_0(z) := 1, \\ q_k(z) := \prod\limits_{\substack{j=1\\ j\ne k}}^{l} \dfrac{z-\varsigma_{jl}}{\varsigma_{kl}-\varsigma_{jl}} \prod\limits_{j=1}^{l-1} \dfrac{\varsigma_{kl}-w_{jl}}{z-w_{jl}}, \quad 1 \le k < l, \end{cases} \tag{4.3}$$

$$\begin{cases} V_0(z,\lambda) := (Vq_0)(z), \\ V_{2k-1}(z,\lambda) := (Vq_k)(z), \quad 1 \le k < l, \\ V_{2k}(z,\lambda) := (V(iq_k))(z), \quad 1 \le k < l. \end{cases} \tag{4.4}$$

We seek to determine a linear combination

$$u_n(z) := \sum_{k=0}^{n-1} a_k V_k(z,\lambda), \quad n = 2l-1, \tag{4.5}$$

which approximately satisfies (1.1) by boundary collocation. The ***collocation points*** are chosen as the generalized Fejér points

$$z_j := \hat{\psi}(e^{2\pi i(j-1)/n}), \quad 1 \le j \le n. \tag{4.6}$$

The function space (4.2) and basis (4.3) have been used to approximate analytic functions on pronouncedly nonconvex regions in [Re2], where the rate of convergence is also discussed. In [Re2] we also used basis (4.5), with $\lambda=0$, and collocation points (4.6) to approximate harmonic functions. The basis (4.3) was shown to be fairly well-conditioned, and computations with basis $\{V_j(z,0)\}_{j=0}^{n-1}$ indicated it is quite well-conditioned, also. In the present application, the basis has to be well-conditioned enough so that we can detect approximate eigenvalues by the method described in Section 2.

The quantities w_{jl}, ς_{jl} and z_j can be computed without explicitly determining $\hat{\psi}$. Assume $|\frac{\partial\hat{\psi}}{\partial n}|$ is known on $\partial\Omega$, where $\frac{\partial\hat{\psi}}{\partial n}$ denotes the outward normal derivative of $\hat{\psi}$ on $\partial\Omega$. Since $\hat{\psi}(\partial\Omega)$ is known, ∂S can be determined by solving an initial value problem for the Cauchy-Riemann equations [Re2]. This initial value problem is ill-posed, but we only need a solution of low accuracy. In the numerical experiments of this section we chose $|\frac{\partial\hat{\psi}}{\partial n}|$ constant on $\partial\Omega$, i.e., we distribute the ς_{jl} and z_j equidistantly with respect to arc length. We also distribute auxiliary nodes $\{z_{jl}\}_{j=1}^{l-1}$ on $\partial\Omega$ equidistantly with respect to arc length, see Figures 4.1-4.2. An initial value problem for the Cauchy-Riemann equations is solved by Euler's method and Figures 4.1-4.2 show approximate stream lines emanating from the z_{jl}. We allocate the w_{jl} on these stream lines on an approximate level curve Γ. Details about the numerical integration method are given in [Re2], [Re4].

We chose $z_0 = z_{1l}$ in (4.1) and evaluated the basis (4.4) for $z \in \partial\Omega$ by integrating along $\partial\Omega$ using a 10-point Lobatto rule between each pair of adjacent collocation points. To compute the double integral in (3.1) we used a two-dimensional midpoint rule defined by

$$\iint_{\Omega} u^2 dxdy \approx h^2 \sum_{(j,k)\in I} u^2(jh,kh) \tag{4.7}$$

where

$$I := \left\{(j,k) : \{(x,y) : |jh-x| < \frac{h}{2},\ |kh-y| < \frac{h}{2}\} \subseteq \Omega \cup \partial\Omega\right\}.$$

We halved h until the value of the integral (4.7) was likely to be known with two significant digits. We point out that the computation of error bounds for (4.5) is very time consuming and therefore generally impractical. We therefore omit further details.

Example 4.1. Let Ω be the interior of the curve

$$\partial\Omega := \{x(t)+iy(t),\ x(t) := 1.4\cos(t)+2.8\cos(2t)-2.8,$$
$$y(t) := 1.6\sin(t)+2.45\sin(t-0.2)+0.98\sin(2t)-0.49\sin(4t)-0.74,\ \ 0 \le t < 2\pi\},$$

see Figure 4.1. The lowest eigenvalue has been computed using expansion (4.5) with different numbers of basis functions.

n	λ	ϵ	upper bound lower bound
49	1.22980	1.4	—
65	1.2305134	$8.4 \cdot 10^{-3}$	1.2^{5}_{0}

Table 4.1

The points w_{jn} for the different bases are shown in Figures 4.1-4.2.

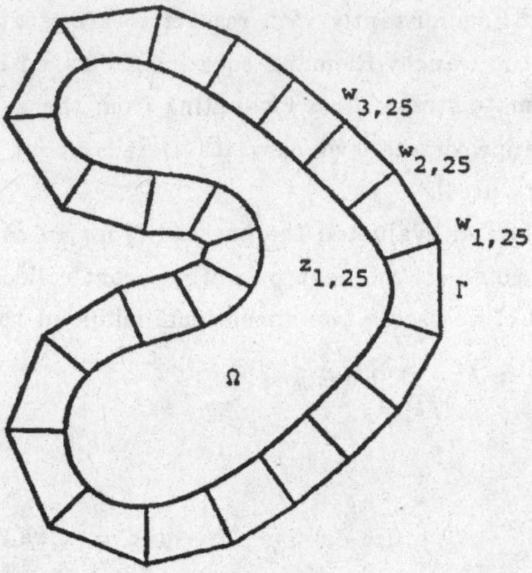

Figure 4.1: 49 basis functions

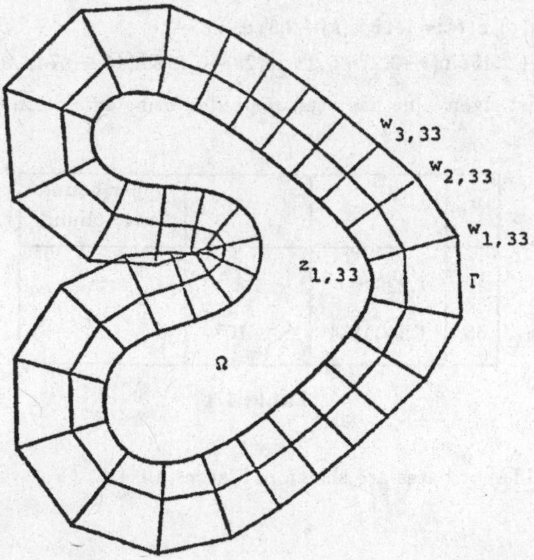

Figure 4.2: 65 basis functions

Acknowledgement

I would like to thank W.J. Cody for providing excellent routines for the evaluation of Bessel functions.

REFERENCES

[BN] R.H.T. Bates and F.L. Ng, Point matching computation of transverse resonances, Int. J. Numer. Meth. Engng., 6 (1973), 155-168.

[CH] R. Courant and D. Hilbert, Methods of Mathematical Physics, vol. 1, Wiley, New York, 1953.

[Cu] J.H. Curtiss, Transfinite diameter and harmonic polynomial interpolation, J. d'Analyse Math., 22 (1969), 371-389.

[Ei] S.C. Eisenstat, On the rate of convergence of the Bergman-Vekua method for the numerical solution of elliptic boundary value problems, SIAM J. Numer. Anal., 11 (1974), 654-680.

[FHM] L. Fox, P. Henrici and C. Moler, Approximation and bounds for eigenvalues of elliptic operators, SIAM J. Numer. Anal, 4 (1967), 89-102.

[Fi] B.E. Fischer, Approximationssätze für Lösungen der Helmholtzgleichung und ihre Anwendung auf die Berechnung von Eigenwerten, Ph.D. thesis, ETH, Zürich, 1983.

[Ga] D. Gaier, Vorlesungen über Approximation im Komplexen, Birkhäuser, Basel, 1980.

[Gu] M.H. Gutknecht, Numerical conformal mapping methods based on function conjugation, J. Comput. Appl. Math., 14 (1986), 31-77, in [T].

[He] P. Henrici, Applied and Computational Complex Analysis, vol. 3, Wiley, New York, 1986.

[HZ] R. Hettich and P. Zenke, Two case-studies in parametric semi-infinite programming, in Systems and Optimization, eds. A. Bagchi and H. Th. Jongen, Lecture Notes in Control and Information Sciences, No. 66, Springer, Berlin, 1985, 132-155.

[KS] J.R. Kuttler and V.G. Sigillito, Eigenvalues of the Laplacian in two dimensions, SIAM Rev., 26 (1984), 163-193.

[Mo] C.B. Moler, Accurate bounds for the eigenvalues of the Laplacian and applications to rhombical domains, Report # CS 121, Computer Science Department, Stanford University, 1969.

[MP] C.B. Moler and L.E. Payne, Bounds for eigenvalues and eigenvectors of symmetric operators, SIAM J. Numer. Anal., 5 (1968), 64-70.

[Re1] L. Reichel, On the computation of eigenvalues of the Laplacian by the boundary collocation method, in Approximation Theory V, eds. C.K. Chui et al., Academic Press, Boston, 1986, pp. 539-543.

[Re2] L. Reichel, On complex rational approximation by interpolation at preselected nodes, Complex Variables: Theory and Appl., 4 (1984), 63-87.

[Re3] L. Reichel, On the determination of boundary collocation points for solving some problems for the Laplace operator, *J. Comput. Appl. Math.*, 11 (1984), 173-196.

[Re4] L. Reichel, Numerical methods for analytic continuation and mesh generation, *Constr. Approx.*, 2 (1986), 23-39.

[SH] B.E. Spielman and R.F. Harrington, Waveguides of arbitrary cross section by solution of a nonlinear integral eigenvalue equation, IEEE Trans. Microwave Theory Techn., MTT-20 (1972), 578-585.

[T] L.N. Trefethen, ed., Numerical Conformal Mapping, *J. Comput. Appl. Math.*, 14 (1986).

[Ve] I.N. Vekua, New Methods for Solving Elliptic Equations, North-Holland, Amsterdam, 1967.

ON THE GEOMETRY OF REAL POLYNOMIALS

Boris Shekhtman
Institute for Constructive Mathematics
Department of Mathematics
University of South Florida
Tampa, FL 33620

Abstract In this article we study the structure of the space of polynomials of degree n, considered as a finite-dimensional Banach space. We give some estimates of the Banach-Mazur distance of this space and its subspaces to the classical Banach spaces $\ell_p^{(m)}$. In the case when the degree of the polynomials is 2^n we show that the space can be decomposed as the direct sum of n subspaces each of which is isometric to a finite dimensional ℓ_∞ space.

1. Introduction.

In this paper we study some geometric properties of the Banach space Π_n that consists of polynomials of degree $n-1$, and equipped with the uniform norm on the interval $[-1,1]$.

Since we are only concerned with the isometric properties of this space we may as well think of Π_n as a set of n-tuples $(a_0,\ldots,a_{n-1})$ with the norm

$$\|(a_0,\ldots,a_{n-1})\| = \sup_{t\in[-1,1]} \left\|\sum_{j=0}^{n-1} a_j t^j\right\|$$

or as a space of n-tuples $(\alpha_0,\ldots,\alpha_{n-1})$ with the norm

$$\|(\alpha_0,\ldots,\alpha_{n-1})\| = \sup_{t\in[-1,1]} \left\|\sum_{0}^{n-1} \alpha_j T_j\right\|$$

when T_j are Chebyshev polynomials. Each of the spaces described above is isometric to Π_n.

We are interested in how "close" the space Π_n (or some isometric copy of it) lies to one of the classical and simple Banach spaces: $\ell_p^{(n)}$

where

$$\ell_p^{(n)} = \{(\alpha_0,\dots,\alpha_{n-1})\} \text{ with the norms}$$

$$\|(\alpha_0,\dots,\alpha_{n-1})\|_p = \begin{cases} (\Sigma|\alpha_j|^p)^{\frac{1}{p}} & \text{for } 1 \le p < \infty \\ \max_j |\alpha_j| & \text{for } p = \infty . \end{cases}$$

It turns out that in some sense the space Π_n relates most closely to $\ell_\infty^{(n)}$ namely it is $\log n$ away from that space.

In a geometric interpretation we are interested in determining the shape of the unit ball of Π_n. The previous remark means for instance that the Hausdorff distance between the unit ball of Π_n and an n-dimensional cube grows as $\log n$. We will need some definitions and propositions about n-dimensional Banach spaces.

Given two n-dimensional Banach spaces E and F the distance $d(E,F)$ is defined to be

$$d(E,F) = \inf\{\|T\|\,\|T^{-1}\|: T \text{ is an isomorphism from } E \text{ onto } F\}.$$

Geometrically $d(E,F)$ is the Hausdorff distance between the unit balls of E and F.

To say that $d(E,F) \le C$, (C being a given constant) is equivalent to exhibiting two bases $\ell_1,\dots,\ell_n$ in E and $f_1,\dots,f_n$ in F and two constants C_1 and C_2 such that the inequalities

$$\text{(1.0)} \qquad \frac{1}{C_2}\|\Sigma\alpha_j f_j\|_F \le \|\Sigma\alpha_j \ell_j\|_E \le C_1\|\Sigma\alpha_j f_j\|_F$$

hold for any choice of scalars $\alpha_1,\dots,\alpha_n$ and $C_1C_2 \le C$. The second notion we will need is that of the projection constant:

Let E be an n-dimensional subspace of a Banach space F then

$$\lambda(E,F) := \inf\{\|P\|: P \text{ is a projection from } F \text{ onto } E\}$$

$$\lambda(E) = \sup\{\lambda(E,F): \ F \supset E_n\}.$$

There are some well known properties that connect the notions of projection constants and the distance:

(1.1) a) $1 \leq \lambda(E) \leq d(E, \ell_\infty^{(n)})$ if $\dim E = n$

(1.2) b) $\lambda(E) \leq d(E,F)\lambda(F)$ if $\dim E = \dim F$

(1.3) c) For any space F of the form $C(K)$ or $\ell_\infty(\Gamma)$, we have $\lambda(E,F) = \lambda(E)$.

Since the space $\ell_\infty^{(n)}$ will play a central role in this paper we need a way to recognize $\ell_\infty^{(n)}$ when we see one. Here is a simple

Proposition 1 (cf[4]). Let E be an n-dimensional subspace in $C(K)$. Then E is isometric to $\ell_\infty^{(n)}$ $(d(E_n, \ell_\infty^{(n)}) = 1)$ iff E has a set of functions $\ell_1(t),\dots,\ell_n(t)$ with the properties $0 \leq \ell_i(t) \leq 1$, $\|\ell_i\| = 1$, $i = 1,\dots,n$; and $\Sigma\ell_i(t) \equiv 1$, $\forall t \in K$.

Throughout this paper we commit the notational abuse of using the letter C with subscripts to denote constants that are not necessarily the same, but do not depend on the dimension.

2. $d(\Pi_n, \ell_\infty^{(n)})$.

In this section we will return to the space of polynomials Π_n and give a simple prove that $d(\Pi_n, \ell_\infty^{(n)}) \sim \log n$.

Geometrically it means that there exists an appropriate isometric copy of Π_n such that the distance from its unit ball to the cube is asymptotically equal to $\log n$.

Theorem 1: For the space Π_n we have

$$\frac{1}{8\sqrt{\pi}} \ell n n \leq d(\Pi_n, \ell_\infty^{(n)}) \leq \frac{4}{\pi} \ell n n + 8$$

Proof: It is well known (cf [5], p.677) that

$$\lambda(\Pi_n, C_{[-1,1]}) \geq \frac{1}{8\sqrt{\pi}} \ell n n.$$

Applying (1.1) and (1.3) we get the left hand side inequality. For the other inequality we need a projection P_n from $C_{[-1,1]}$ onto Π_n that interpolates at the Chebyshev roots: $\left\{\tau_k^{(n)} = \cos\frac{2k-1}{2n}\pi,\ k = 0,\dots,n-1\right\}$. The norm of this projection satisfies $\|P_n\| \leq \frac{4}{\pi} \ell n n + 8$ (cf [5], p.542). We can write this projection in the form

$$P_n f = \sum_{k=0}^{n-1} f(\tau_k^{(n)})\varphi_k$$

where φ_k are the fundamental Lagrange polynomials with respect to $\{\tau_k^{(n)}\}$.

For arbitrary $\alpha = (\alpha_1,\dots,\alpha_n)$ we have

$$\|\alpha\|_\infty = \max_j |\alpha_j| = \max_j |(\Sigma\alpha_k\varphi_k)(\tau_j^{(n)})| \leq \|\Sigma\alpha_k\varphi_k\|.$$

On the other hand, let f be a function whose norm is $\max|\alpha_j|$ such that $f(\tau_j^{(n)}) = \alpha_j$, then

$$\|\Sigma\alpha_k\varphi_k\| = \|P_n f\| \leq \|P_n\| \|f\| = \|P_n\|\cdot\max|\alpha_k|.$$

All together we have

$$\|\alpha\|_\infty \leq \|\Sigma\alpha_k\varphi_k\| \leq (\frac{4}{\pi} \ell n n + 8)\|(\alpha\|_\infty)$$

which is just the needed estimate of type (1.0) □

Remark 1. We see from (1.0) that the Banach-Mazur distance is connected with the choice of isomorphism or equivalently with a choice of basis in the space. The proof of the above theorem shows that the basis

$$\varphi_k^{(n)}(t) = \frac{T_{n-1}(t)}{(t - \tau_k^{(n)})T'_{n-1}(\tau_k^{(n)})} \qquad (0 \leq k \leq n-1)$$

is the closest (up to a constant) basis to the canonical vector basis in $\ell_\infty^{(n)}$.

In the next section we will exhibit another basis with this property.

Remark 2. Clearly Π_n is not the only Banach space with the property that $d(\Pi_n, \ell_\infty^{(n)}) \sim \log n$. For instance a classical $\ell_{p_n}^{(n)}$ space with

$$p_n = \frac{2\,\ell n(n)}{\ell n(n) - \ell n(\ell n(n))}$$

behaves in a similar manner.

3. Partial and complete decomposition of Π_n.

The first result in this section is a reformulation of the Hermitt-Fejer interpolation. It shows that Π_{2n} contains an n-dimensional subspace which is isometric to $\ell_\infty^{(n)}$ or equivalently one can find an n-dimensional hyperspace in $\mathbb{R}_{2n}$ such that its intersection with the unit ball of Π_{2n} is a perfect n-dimensional cube. (This result was also proved in [1]).

Based on this procedure we show that Π_n can be decomposed.

$$\Pi_{2^m} = E_0 \oplus E_1 \oplus E_2 \oplus E_4 \ldots \oplus E_{2^{m-1}}$$

and each E_j is isometric to $\ell_\infty^{(\dim E_j)}$.

Theorem 2 (Fejer Decomposition). For any $n > 0$ we have

$$\Pi_{2n} = E_n \oplus F_n$$

where $E_n = \operatorname{span}\left\{\left[\frac{T_n(t)}{n(t - \tau_k^{(n)})}\right]^2 (1 - t\cdot\tau_k^{(n)})\right\}_0^{n-1}$ is isometric to $\ell_\infty^{(n)}$; $F_n = T_n \cdot \Pi_n$.

Proof: Every polynomial $p \in \Pi_{2n}$ can be uniquely represented as

$$p(t) = \sum_{k=1}^{n} p(\tau_k^{(n)}) A_k^{(n)}(t) + \sum_k p'(\tau_k^{(n)}) B_k^{(n)}(t) \tag{3.0}$$

where

$$A_k^{(n)}(t) = \left[\frac{T_n(t)}{n(t - \tau_k^{(n)})}\right]^2 (1 - t\cdot\tau_k^{(n)}) \tag{3.1}$$

$$B_k^{(n)}(t) = \left[\frac{T_n(t)}{n(t - \tau_k^{(n)})}\right]^2 (1 - (\tau_k^n)^2)(t - \tau_k^{(n)}) = \tag{3.2}$$

$$= T_n(t)\cdot\left\{\frac{T_n(t)}{n(t - \tau_k^{(n)})} (1 - (\tau_k^{(n)})^2)\right\}.$$

Letting $E_n = \mathrm{span}\{A_k^{(n)}\}$ and noticing that in the braces we have the span of fundamental Lagrange polynomials with respect to the Chebyshev points we have

$$\Pi_{2n} = E_n \oplus T_n\cdot\Pi_n.$$

It remains to prove that E_n is isometric to $\ell_\infty^{(n)}$. Clearly

a) $A_k^{(n)}(t) \geq 0$

b) $A_k^{(n)}(t) = 0$ iff $t = \tau_j^{(n)}$, $j \neq k$; $A_k^{(n)}(\tau_k^{(n)}) = 1$

c) $\Sigma A_k^{(n)}(t) \equiv 1$ (Letting $p(t) \equiv 1$ in (3.0)).

Using Proposition 1 we see that E_n is isometric to $\ell_\infty^{(n)}$, more over $A_k^{(n)}$ form a basis which is equivalent to the canonical vector basis in $\ell_\infty^{(n)}$ i.e.

$$\|\Sigma\alpha_k A_k^{(n)}\| = \max_k |\alpha_k| = \max_k |(\Sigma\alpha_j A_j^{(n)})(\tau_k^{(n)})| \tag{3.4}$$ □

It is natural to write now

$$\Pi_{2^n} = E_{2^{n-1}} \oplus T_{2^n} \Pi_{2^{n-1}} = E_{n^{n-1}} \oplus T_{2^n}\left[E_{2^{n-2}} \oplus T_{2^{n-1}} \Pi_{2^{n-2}}\right] =$$

$$E_{2^{n-1}} \oplus T_{2^n} E_{2^{n-2}} \oplus T_{2^n} \cdot T_{2^{n-1}} \Pi_{2^{n-2}} = \dots$$

to obtain a complete decomposition. Which brings us to the next

<u>Theorem 3</u>. For the space Π_{2^n} we have decomposition

$$\Pi_{2^n} = E_0 \oplus E_1 \oplus \dots \oplus E_n$$

where E_j is isometric to the space $\ell_\infty^{2^{j-1}}$, $j = 1,\dots n$. E_0 is isometric to ℓ_∞^1.

<u>Proof</u>. It follows from the previous theorem that

$$\Pi_{2^n} = \sum_{j-0}^{n} \oplus E_j \text{ where } E_n = \text{span}\left\{\left[T_{2^n}(t)/n(t - \tau_k^{(2^n)})\right]^2 (1 - t\tau_k^{(2^n)})\right\}$$

and

$$E_j = \left[\prod_{\ell=j+1}^{n} T_{2^\ell}\right] \cdot \text{span}\left\{\left[\frac{T_{2^j}(t)}{n(t - \tau_k^{(2^j)})}\right]^2 (1 - t\tau_k^{(2^j)})\right\}.$$

We have already seen that E_n is isometric to $\ell_\infty^{2^{n-1}}$, E_0 being a one-dimensional space is clearly isometric to ℓ_∞^1. It remains to prove that E_j is isometric to $\ell_\infty^{2^{j-1}}$. For that purpose we need to review the properties of the polynomials

$$A_k^{(j)}(t) := \left[\frac{T_{2^j}(t)}{n(t - \tau_k^{(2^j)})}\right]^2 (1 - t\tau_k^{(2^j)})$$

We know from the proof of the Theorem 2 that

$$A_k^{(j)}(t) \geq 0; \quad A_k^{(j)}(\tau_m^{(2^j)}) = \delta_{k,m}, \quad \sum_k A_k^{(j)}(\tau_m^{(2^j)}) \equiv 1$$

and

$$\max|\alpha_m| = \|\Sigma\, \alpha_k A_k^{(j)}(t)\| = \max_m |(\Sigma\, \alpha_k A_k^j)(\tau_m^{2^j})|.$$

Let

$$\varphi_k^{(j)}(t) = \left[\prod_{\ell=j+1}^{n} T_{2^\ell}(t)\right] A_k^{(j)}(t).$$

We have

$$\left\|\Sigma\, \alpha_k \varphi_k^{(j)}(t)\right\| = \left\|\prod_{\ell=j+1}^{n} T_{2^\ell}(t)(\Sigma\, \alpha_k A_k^{(j)}(t))\right\| \leq$$

$$\leq \left\|\Pi\, T_{2^\ell}(t)\right\| \cdot \left\|\Sigma\, \alpha_k A_k^{(j)}(t)\right\| \leq \max|\alpha_k|.$$

To prove the converse inequality let m be such that $|\alpha_m| = \max|\alpha_k|$. Then

$$\left\|\Sigma\, \alpha_k \varphi_k^{(j)}(t)\right\| \geq |(\Sigma\, \alpha_k \varphi_k^{(j)})(\tau_m^{(2^j)})| = |\alpha_m| \cdot |\varphi_k^{(j)}(\tau_m^{(2^j)})| =$$

$$= |\alpha_m| \cdot |\prod_{\ell=j+1}^{n} T_{2^\ell}(\tau_m^{(2^j)})|.$$

To finish off the proof we need to show that

$$|T_{2^\ell}(\tau_m^{(2^j)})| = 1 \quad \text{for} \quad \ell > j.$$

By direct computation

$$T_{2^\ell}(\tau_m^{(2^j)}) = \cos\left[2^\ell \cdot \frac{(2^m - 1)}{2 \cdot 2^j}\, \Pi\right] = \cos((\text{integer}) \cdot \Pi) = \pm 1 \quad \square$$

In the last theorem we exhibited a basis for Π_{2^n} that consists of the set of functions $\varphi_k^{(j)}$ such that every $p(x) \in \Pi_{2^n}$ can be uniquely represented as

$$\sum_{j=0}^{n} \sum_{k=2^{j-2}}^{2^{j-1}} \alpha_k^{(j)} \varphi_k^{(j)}.$$

Problem 1. Is there a similiar formula for Π_n for arbitrary n?

4. On the $d(\Pi_n, \ell_{p(n)}^{(n)})$.

In this section we will examine the distance from polynomials in the uniform norm to other classical Banach spaces: $\ell_p^{(n)}$. In particular using the results of the previous sections we will show that

$$d(\Pi_n, \ell_p^n) \geq C \cdot \sqrt{\log n}$$

i.e. it can not be much better that $d(\Pi_n, \ell_\infty^{(n)})$. In fact we suspect that it is worse.

Proposition 2: Let $\gamma_n = \begin{cases} n^{\frac{1}{p(n)}} & \text{if} \quad p(n) \geq 2 \\ \sqrt{n} & \text{if} \quad p(n) < 2 \end{cases}$. Then there exist constants C_1 and C_2 such that

$$C_2 \frac{\gamma_n}{\log n} \leq d(\Pi_n, \ell_{p(n)}^{(n)}) \leq C_1 \log n \cdot \gamma_n. \tag{4.0}$$

Proof: It is well known (cf [2]) that $\gamma_n \leq d(\ell_{p(n)}, \ell_\infty^{(n)}) \leq (1+\sqrt{2})\gamma_n$. Now using the triangle inequality and Theorem 1 we have

$$d(\Pi_n, \ell_{p(n)}^{(n)}) \leq d(\Pi_n, \ell_\infty^{(n)}) d(\ell_{p(n)}^{(n)}, \ell_\infty^{(n)}) \leq C_1 \log n \; \gamma_n.$$

On the other hand

$$C \gamma_n \leq d(\ell_{p(n)}^{(n)}, \ell_\infty^{(n)}) \leq d(\Pi_n, \ell_{p(n)}^{(n)}) \cdot d(\Pi_n, \ell_\infty^{(n)}) \quad \text{and}$$

$$d(\Pi_n, \ell_{p(n)}^{(n)}) \geq \frac{C\,\gamma_n}{d(\Pi_n, \ell_\infty^{(n)})} \geq C_2 \frac{\gamma_n}{\log n} \qquad \square$$

This estimate shows that if we want $\ell_{p(n)}^{(n)}$ to be close to Π_n we need to choose $p(n) \to \infty$.

Choosing $p(n) = \frac{\log n}{\log 3}$ the Proposition 2 gives

$$(4.1) \qquad C_2 \leq d(\ell_{p(n)}^{(n)}, \Pi_n) \leq \tilde{C}_1 \log n.$$

Choosing $p(n) = \frac{2 \log n}{\log n - \log(\log n)}$ we have (4.1) jointly with $d(\ell_{p(n)}^{(n)}, \ell_\infty^{(n)}) \sim \log n$.

This evidence may suggest that for some $p(n)$: $d(\ell_{p(n)}^{(n)}, \Pi_n)$ is significantly smaller than $d(\Pi_n, \ell_\infty^{(n)})$. However

<u>Theorem 4</u>: For any sequence $p(n)$ we have

$$d(\ell_{p(n)}^{(n)}, \Pi_n) \geq C \cdot \sqrt{\log n}.$$

<u>Proof</u>: In view of Proposition 2 we can choose $p(n) \to \infty$. Let $\Gamma_n := \Gamma(n) = d(\ell_{p(n)}^{(n)}, \Pi_n)$. Then

$$C \cdot \log 2n \leq d(\Pi_{2n}, \ell_\infty^{2n}) \leq d(\ell_{p(n)}^{(2n)}, \ell_\infty^{(2n)}) \cdot d(\ell_{p(2n)}^{(2n)}, \Pi_{2n}) \leq$$

$$\leq C_1 \Gamma_{2n} (2n)^{\frac{1}{p(2n)}}.$$

So

$$(4.1) \qquad (2n)^{\frac{1}{p(2n)}} \geq \frac{C \log 2n}{\Gamma_{2n}}.$$

Let $T: \Pi_{2n} \to \ell_{p(2n)}^{(2n)}$ so that $\|T\| \, \|T^{-1}\| \leq \Gamma_{2n}$ and let E_n be an n-dimensional subspace of Π_{2n} isometric to $\ell_\infty^{(n)}$. (Such subspace exists by theorem 2). Then

$$TE_n =: F_n \subset \ell_{p(2n)}^{(2n)}$$

and

(4.2) $$d(F_n, \ell_\infty^{(n)}) \le \Gamma_{2n}.$$

Since F_n is a subspace of $\ell_{p(2n)}^{(2n)}$ we have an estimate (cf [3])

$$d(F_n, \ell_2^{(n)}) \le n^{\frac{1}{2} - \frac{1}{p(2n)}}$$

and by triangle inequality

$$d(F_n, \ell_\infty^{(n)}) \ge d(\ell_2^{(n)}, \ell_\infty^{(n)})/d(F_n, \ell_2^{(n)}) \ge \frac{n^{\frac{1}{2}}}{n^{\frac{1}{2} - \frac{1}{p(2n)}}} = n^{\frac{1}{p(2n)}}.$$

Together with (4.2) and (4.1)

$$\Gamma_{2n} \ge n^{\frac{1}{p(2n)}} \ge C\,\frac{\log 2n}{\Gamma_{2n}}.$$

The conclusion is

$$(\Gamma_{2n})^2 \ge C\log(2n) \quad \text{and} \quad \Gamma_{2n} \ge C\cdot\sqrt{\log 2n}.$$

For Π_{2n+1} the proof is similiar. □

<u>Problem 2</u>: Can the estimate in the theorem 4 be improved to $d(\Pi_n, \ell_{p(n)}^{(n)}) \ge C\cdot\log n$?

<u>5. Enveloping</u> Π_n.

In the previous section we dealt with a possibility of embedding $\ell_\infty^{(n)}$ into Π_n for $m < n$. In this section we deal with the inverse embedding

$$\Pi_n \subset \ell_\infty^{(m)}, \; m > n.$$

We show that such embedding does not exist for any $m \ge n$. We also will discuss a number $\alpha > 1$ such that $d(\ell_\infty^{(m)}, F) = \alpha$ and $\Pi_n \subset F$.

Definition: A finite-dimensional Banach space E is called a polyhedral space if it is isometric to a subspace of $\ell_\infty^{(m)}$ for some integer m.

Geometrically E is a polyhedral space iff the unit ball of E is a convex polyhedron. Our next theorem shows that Π_n is not a polyhedral space and hence Π_n can not be embedded into $\ell_\infty^{(m)}$ for any m. In fact we will prove a slightly more general result.

Theorem 5. Let E_n be an n-dimensional subspace of $C[a,b]$ and E_n contains a Chebyshev system $\varphi_1, \varphi_2, \dots, \varphi_k$ for some $k > 2$. Then E_n is not a polyhedral spacc.

Proof: Let E_n be a polyhedral space. Then its unit ball has finitely many extreme points and hence the Choquet boundary of E_n is finite. On the other hand E_n contains a Chebyshev system $\varphi_1, \dots \varphi_k$ $(k \geq 3)$. From well-known properties of Chebyshev systems, for any $z \in (a,b)$ there exists a function

$$\varphi(t) = \sum_{j=1}^{k} \alpha_j \varphi_j$$

such that $\varphi(t) < 0$ for $t \in [a,b]\setminus\{z\}$ and $\varphi(z) = 0$. It is also well-known that the span of Chebyshev systems contains a strictly positive function. Hence (cf [6]) the Choquet boundary of E_n contains (a,b) and in particular is uncountable. □

Remark. The theorem fails if E_n contains Chebyshev system of order 2. For instance let $E_2 = \text{span}\{1,t\}$ on $[-1,1]$. Then $\varphi_1(t) = \frac{1}{2} - \frac{1}{2}t$, $\varphi_2(t) = \frac{1}{2} + \frac{1}{2}t$ are functions in E_2 and satisfy the conditions of Proposition 1. Therefore

$$E_2 \text{ isometric to } \ell_\infty^{(2)}.$$

We now turn to the question of how far away a given subspace F has to be from ℓ_∞^m to contain Π_n. An easy answer to that question is provided by the following

<u>Theorem 6</u> (cf[4]). Let E be an n-dimensional subspace of $C_{[a,b]}$. Then for any $\epsilon > 1$ there exists an integer m and a subspace $F \subset C_{[a,b]}$ with $\dim F = m$ such that

$$d(F, \ell_\infty^{(m)}) < \epsilon$$

and

$$E \subset F.$$

In view of this theorem the following question may be of interest.

Let $\alpha(n, \epsilon)$ be the smallest integer m that satisfies the conclusion of the last theorem.

<u>Problem 3</u>. Find good estimates for $\alpha(n,\epsilon)$.

Equivalently for given n,m let $\beta(n,m)$ be the smallest of the numbers β such that there exists a subspace $F \supset \Pi_n$ with

$$\dim F = m \text{ and } d(F, \ell_\infty^{(m)}) \leq \beta.$$

Clearly $\beta(n,n) \sim \log n$. For small $m - n$ (such as $m - n \sim O(\log^2 n)$) we can generalize this result:

<u>Theorem</u>: For all $m \geq n$; $\beta(n,m) \geq \dfrac{\log n}{8\sqrt{m-n+1}}$.

<u>Proof</u>: Let F be of dimension m and contains Π_n. Then Π_n is a subspace in F of codimension $m - n$ and hence cf[1] $\lambda(\Pi_n, F) \leq \sqrt{m-n+1}$.

Clearly

$$\lambda(\Pi_n) \leq \lambda(\Pi_n, F) \cdot \lambda(F).$$

Thus $\lambda(F) \geq \frac{\lambda(\Pi_n)}{\lambda(\Pi_n, F)} \geq \frac{1}{8\sqrt{\pi}} \frac{\log n}{\sqrt{m-n+1}}$.

To obtain the conclusion of the theorem observe that

$$\frac{1}{8\sqrt{\pi}} \frac{\log n}{\sqrt{m-n+1}} \leq \lambda(F) \leq d(F, \ell_\infty^{(m)}) \leq \beta(n,m) \qquad \square$$

We conclude this section with an application of the Theorem 2 to a result of Theorem 6 type.

<u>Proposition 3</u>. Let E be a finite-dimensional subspace of $C_{[-1,1]}$. The $\forall \epsilon > 0$ there exists a <u>polyhedral</u> space F so that $d(F,E) < \epsilon$.

<u>Proof</u>: Let E_m be as in the Theorem 2, and let

$A_k^{(m)}(t) = \left[\frac{T_m(t)}{n(t-z_k^{(m)})}\right]^2 (1 - t\tau_k^{(n)})$ be the polynomials that span E_m.

It is well known (cf [5], p.554) that for any $\epsilon > 0$ and any $f \in C_{[-1,1]}$ there is an M so that

$$\|f - e_m(f)\| < \epsilon, \quad \text{for some } e_m(f) \in E_m, \quad \text{for all } m > M.$$

Let $E = \text{span}\{x_1, \ldots, x_n\}$. Then for any $\epsilon > 0 \ \exists\ m$ so that $\|e_m(x_i) - x_i\| < \epsilon$. Then for the operator T defined by $Tx_i = e_m(x_i)$ we have $T: E \to \text{span}\{e_m(x_i)\} =: F$ and $\|T\|\,\|T^{-1}\| \leq 1 + \epsilon$. Hence $F \subset E_m$ and thus polyhedral. $\square$

<u>References</u>

[1] E.W. Cheney and K.H. Price, Minimal projections, In Approximation Theory (A.Talbot ed.)(1970), pp. 261-289.

[2] V.I. Gurarii, M.E. Kadec and V.I. Macaev, On the distance between isomorphic L_p spaces of finite dimension, Mat.Sb. (112) 4(1966), pp. 481-489.

[3] D.R. Lewis, Finite dimensional subspaces of L_p, Studia Math. 63(1978), pp. 207-212.

[4] E. Michael and A. Pelczynski, Separable Banach spaces that admit ℓ_n^∞-approximation, Israel J.Math vol.4(1966), pp. 189-198.

[5] I.P. Natanson, Constructive theory of approximation, Moscow 1949.

[6] R.R. Phelps, Lectures on Choquet's theorem, D.Van Nostrand Co, 1966.

A NOTE ON A THEOREM BY H. N. MHASKAR AND E. B. SAFF

H. R. Stahl
TU-Berlin/Sekr. 6-8
Franklinstr. 28/29
1000 Berlin 10, West Germany

Abstract. In this note it is shown that the asymptotic error estimate for weighted Chebyshev polynomials holds true under the same assumptions as those normaly required for the classical non-weighted case.

In [1] Mhaskar and Saff have investigated the asymptotic behavior of weighted Chebyshev polynomials. In this note we improve the asymptotic error estimate given in [1, Thm. 2.2 and Thm. 4.2]. Essentially, we shall show that the technical assumptions used in the quoted theorems are avoidable.

We shall use the notation of [1]. The errors of the weighted Chebyshev polynomials are defined as

(1) $$E_n(w) := \inf\{\|w(x)^n[x^n - p_{n-1}(x)]\|_\Sigma;\ p_{n-1} \in \Pi_{n-1}\},\quad n \in \mathbb{N},$$

where Π_n is the set of polynomials of degree at most $n \in \mathbb{N}$, $\|\cdot\|_A$ the sup norm on A, and $w(x) \geq 0$, $x \in \mathbb{R}$, a weight function having the following four properties:

(i) $\Sigma := \operatorname{supp}(w)$ is of positive capacity.
(ii) The restriction of w to Σ is continuous on Σ.
(iii) The set $Z := \{x \in \Sigma;\ w(x) = 0\}$ has capacity zero.
(iv) If Σ is unbounded then $|z|w(x) \to 0$ as $|x| \to \infty$, $x \in \Sigma$.

In [1] it has been proved that there uniquely exists a compact set $S_w \subset \Sigma\setminus Z$ that maximizes the functional

(2) $$F(K) := \log C(K) - \int Q(x)d\nu_K(x)$$

and has the property that $F(K) = F(S_w)$ implies $S_w \subset K$ for any compact set $K \subset \Sigma \setminus z$ of positive capacity. In (2) $Q(x) := -\log w(x)$, $C(K)$ is the (logarithmic) capacity of K, and ν_K the unit equilibrium distribution on K. Using results from [1] we prove the following theorem:

<u>THEOREM</u>-- <u>We have</u>

$$(3) \qquad \lim_{n\to\infty} E_n(w)^{\frac{1}{n}} = e^{F(S_w)} .$$

Since it has been proved in [1, Thm. 2.1 (d)] that $E_n(w)^{\frac{1}{n}} \geq \exp(F(S_w))$ for all $n \in N$, it remains only to be shown that

$$(4) \qquad \limsup_{n\to\infty} E_n(w)^{\frac{1}{n}} \leq e^{F(S_w)}$$

holds without the special assumptions made in [1, Thm. 2.2] or in the remark following this theorem. Thus, we get an asymptotic error estimate of the same generality as in the classical case $w(x) \equiv 1$, where (3) holds for arbitrary compact sets $\Sigma \subset \mathbb{R}$. In the classical case we have $C(\Sigma \setminus S_w) = 0$, $F(S_w) = F(\Sigma) = \log C(\Sigma)$.

<u>Proof</u>: From [1, Thm. 2.3(f) and (d)] we get

$$(5) \qquad \int \log|x - t| d\mu_w(t) \leq Q(x) + F(S_w)$$

for x q.e. on Σ, where μ_w is a probability measure defined in [1, Thm. 2.3(b)] by a minimal energy property. We have $S_w = \operatorname{supp}(\mu_w)$. Let Σ_1 denote the set of points $x \in \Sigma$ for which (5) holds and $\Sigma_o := \Sigma \setminus \Sigma_1$. Hence, $C(\Sigma_o) = 0$. We shall derive (4) from (5) after locating some extra zeros of the weighted polynomials at points of Σ_o. This procedure will ensure that the deficiency of the inequality (5) will not spoil our upper estimate for the asymptotic errors $E_n(w)$.

Since $S_w = \operatorname{supp} \mu_w$ is compact, it follows from property (iv) and (5) that Σ_0 is bounded. As $C(\Sigma_0) = 0$, there exists a probability measure v_0 with compact support in $\mathbb{R}$ and

$$(6) \qquad p_0(x) := \int \log|x - t| dv_0(t) = -\infty$$

for $x \in \Sigma_0$. Again from property (iv) it follows that $c_0 := \sup_{x \in \Sigma} [p_0(x) - Q(x)] < \infty$.

Let $0 < \alpha < 1$ be arbitrary, and define $\mu_\alpha := (1-\alpha)\mu_w + \alpha v_0$. Since $Q(x)$ is bounded from below on Σ, it follows from (5) and (6) that

$$(7) \qquad \int \log|x - t| d\mu_\alpha(t) \leq Q(x) + (1 - \alpha)F(S_w) + \alpha c_0$$

for all $x \in \Sigma$. For any $n \in N$ we can select n points $t_{kn} \in \Sigma$, $k = 1,\ldots,n$, such that the sequence of counting measures v_n defined by

$$(8) \qquad v_n(B) := \operatorname{card}\{k, \ t_{kn} \in B, \ k=1,\ldots,n\}$$

for any real set $B \subset \mathbb{R}$ has the weak limit

$$(9) \qquad \lim_{n\to\infty} \frac{1}{n} v_n = \mu_\alpha .$$

This selection is possible since the discrete measures are dense in the unit ball of positive measures. From (9) with the so-called principle of descent in potential theory [2, Thm. 1.3] it follows that

$$(10) \qquad \limsup_{n\to\infty} \int \log|x - t| dv_n(t) \leq \int \log|x - t| d\mu_\alpha(t)$$

uniformly in x on every compact set in $\mathbb{R}$ on which the right-hand side is bounded from below. Using $G_n(x) := \Pi_{k=1}^{n}(x - t_{kn})$, (7), (10), and property (iv) yields

$$(11) \qquad \limsup_{n\to\infty} \|w(x)^n G_n(x)\|_{\Sigma}^{\frac{1}{n}} \leq e^{\alpha c_o}[e^{F(S_w)}]^{1-\alpha} .$$

As $\alpha > 0$ was arbitrary, (11) implies (4) which together with [1, Thm. 2.1(d)] completes the proof.

REFERENCES

1. H. N. Mhaskar and E. B. Saff (1985): Where does the sup norm of a weighted polynomial live? Constr. Approx., 1, 71-91.

2. N. S. Landkof (1972): Foundations of Modern Potential Theory. Berlin: Springer-Verlag.

EXISTENCE AND UNIQUENESS OF RATIONAL INTERPOLANTS WITH FREE AND PRESCRIBED POLES

Herbert Stahl*
TU-Berlin/Sekr. FR 6-8
Franklinstr 28/29
1000 Berlin 10
Fed. Rep. Germany

Abstract. We study existence and uniqueness problems associated with different definitions of rational interpolants with free or prescribed poles. The treatment of the subject is rather general; it includes interpolation at infinity, at confluent points, and interpolation in polar singularities.

1. Introduction

We shall study different definitions of rational interpolants. Unlike the situation in the theory of polynomial interpolation, there does not exist a unique definition that could be considered as the only natural or the most general one. Instead, there exist quite different possibilities, all have their specific advantages and disadvantages. One major distinction is between rational interpolants with free versus rational interpolants with prescribed poles, but there are further aspects that have to be considered. In order to show how complicated the subject can become we mention that even the distinction between rational interpolants with free versus prescribed poles can be bridged by a definition of rational interpolants with only partly prescribed poles. (See section 8 below.)

For all definitions considered here we investigate existence and uniqueness. For multipoint Padé approximants we also study the possibility of interpolation defects, and as a result get sufficient conditions for the solvability of the Cauchy-Jacobi interpolation problem.

---

*) The paper was written while the author was visiting the Institute for Constructive Mathematics at the University of South Florida, Tampa.

The material of the paper, as far as interpolants with free poles are concerned, is similar to that of a paper by MEINGUET (1970); however, there are two major difference: First, the problem is treated in greater generality. Unlike Meinguet we include interpolation at confluent (i.e. indentical) points (which is known as Hermite interpolation in the polynomial case), interpolation at infinity, and interpolation at polar singularities. The last two topics mark a specific difference between polynomial and rational interpolation. In a certain sense it is natural to extend the definitions of rational interpolation to all points of the extended complex plane $\bar{\mathbb{C}}$. This can be done both with respect to interpolation points as well as with respect to the values of the function to be interpolated. Although the extension can be considered as being natural, it nevertheless demands some care.

The second major difference is related to the analytic instruments applied. Meinguet's treatment of the problem is mainly based on determinants, using such beautiful classical tools as bigradients and Sylverster's identity. Here, we shall avoid, as a general policy, the use of determinants. Thus, we do not have so many connections with classical algebra and the classical theory of determinants, but on the other hand have a somewhat simpler and at the same time more general, and may be, also more unified method for proving the different existence and uniqueness results.

The basic defintions of the subject are very classical. They go back to Lagrange, CAUCHY (1821), JACOBI (1846), and KRONECKER (1881) (for a historic survey including the pre-Cauchy period we refer to BREZINSKI (1979)). It has to be mentioned that despite the constructive character of the arguments the paper is not concerned with numerical algorithms. A good survery of this field can be found in BAKER & GRAVES-MORRIS (1981), Part II, Ch. 1.

The outline of the paper is as follows: In Section 2 we introduce and discuss the basic notation. In Section 3 we shortly touch on the theory of <u>polynomial interpolation</u>, thereby setting the stage for the discussion of the <u>Cauchy-Jacobi interpolation problem</u> in Section 4. The fact that the Cauch-Jacobi interpolation problem is not always solvable leads naturally to the definition of <u>multipoint Padé approximants</u>, also known as <u>modified</u> or <u>linearized rational interpolants</u>. These approximants are defined in Section 5. While the general existence and uniqueness theorem is the definite advantage of multipoint Padé approximants, its draw-back is the possibility of interpolation defects, which are studied in Section 6. As a result we get sufficient conditions for the <u>solvability</u> of the Cauchy-Jacobi interpolation problem. In Section 7 rational interpolants with prescribed poles are defined and discussed. In a certain sense they are an exact analogue of polynomial interpolants. While in Section 2 through Section 7 it has been assumed that the function to be interpolated has finite values at interpolation points and also finite derivatives in case of confluent interpolation, we shall drop this assumption in Section 8, and give an extension of all definitions to functions that may be of polar character at interpolation points. The paper is an extended version of the first part of a talk on rational interpolation and approximation given at the University of South Florida in Spring 1986.

2. Interpolation and Interpolation Schemes

Given a triangular system of interpolation points

$$(2.1)\qquad \begin{array}{l} a_n \\ \ldots \\ a_{n1},\ldots,a_{nn} \\ \ldots \end{array}$$

in $\mathbb{C}$ or $\overline{\mathbb{C}}$, and a function f defined at these points, we are looking for a function g from a specified class (polynomials, rational functions, etc.) that interpolates f in the points $a_{n1},\ldots,a_{nn}$ of a specific row of (2.1). If some of the points a_{nj}, $j = 1,\ldots,n$, are identical, then they will be called confluent interpolation points. In this case interpolation extends to an appropriate number of derivatives of f and g. More precisely: Let $\{a_{n1},\ldots,a_{nn}\}$ contain m $(m \geq 1)$ distinct points $a_1,\ldots,a_m \in \overline{\mathbb{C}}$, each point a_j, $j = 1,\ldots,m$, appearing m_j times in the set $\{a_{n1},\ldots,a_{nn}\}, m_1+\ldots+m_m = n$, and let us first assume $a_j \neq \infty$ for all $j = 1,\ldots,m$. Then interpolation of a given function f in $\{a_{n1},\ldots,a_{nn}\}$ by a function g means that

$$(2.2)\qquad f^{(\ell)}(a_j) = g^{(\ell)}(a_j),\ \ell = 0,\ldots,m_j - 1,\ j = 1,\ldots,m,$$

where the superscript (ℓ) denotes the ℓ-th derivative. If one of the a_j's is equal to ∞, let us say $a_1 = \infty$ and $m_1 > 1$, then we define the derivative at ∞ by mapping infinity onto zero: Put

$\tilde{f}(z) := f\left[\frac{1}{z}\right]$ and $\tilde{g}(z) := g\left[\frac{1}{z}\right]$, then the function f is interpolated in $a_1 = \infty$ by g with an order of contact m_1 if

$$(2.3) \qquad \tilde{f}^{(\ell)}(0) = \tilde{g}^{(\ell)}(0), \ \ell = 0,\ldots,m_1-1.$$

If (2.2), and in case of $a_1 = \infty$ also (2.3), is satisfied then we say that g interpolates f and vice versa in the set $\{a_{n1},\ldots,a_{nn}\}$ in Hermite's sense.

Of course, in order that (2.2) and (2.3) is well defined, we have to ensure that both functions f and g are at least m_j-1 times differentiable at each point a_j of the considered row in (2.1).

Before continuing with the general discussion of interpolation we shall introduce some notation: Any row $A(n) := \{a_{n1},\ldots,a_{nn}\}$, $a_{nj} \in S \subset \overline{\mathbb{C}}$, of (2.1) is called a (n-th order) interpolation set (of S) and $A := (\ldots,A(n),\ldots)$ a interpolation scheme. With any finite set A of points from $\overline{\mathbb{C}}$ we associate the counting measure $\nu(A)$ defined by $\nu(A)(S) := \text{card } (A|_S)$ for every set $S \subset \overline{\mathbb{C}}$, where $A|_S$ is the subset of all points of A contained in S. This subset is also called the restriction of A to S. Since points of $\overline{\mathbb{C}}$ cannot be contained more than once in the set A, it is in general not correct in a strict set-theoretical terminology to write $A \subset \overline{\mathbb{C}}$. We circumvent this difficulty by introducing the support $S(A)$ of an interpolation set A, which is by definition the smallest subset of $\overline{\mathbb{C}}$ containing all points $a \in A$. Thus, $S(A)$ is equal to the support of the counting measure $\nu(A)$ associated with A.

Let Π_n denote the set of all polynomials of degree at most $n \in \mathbb{N}$ with complex coefficients, $\Pi_{mn} := \{P/Q;\ P \in \Pi_m,\ Q \in \Pi_n,\ Q \not\equiv 0\}$, $m,n \in \mathbb{N}$, and $Z(P)$, $P \in \Pi_n$, the set of all zeros of the polynomial P taking account of multiplicities. Thus, card $Z(P) =$ deg P. We call $Z(P)$ the zero set of P.

Every polynomial P is defined by its zero set $Z(P)$ up to a constant factor. In order to get uniqueness we have to introduce a normalization. Let $P_n \subset \Pi_n$, $n \in \mathbb{N}$, denote the set of all polynomials P normalized by the identity

$$(2.4) \qquad P(z) = \prod_{\zeta \in Z(P)} H(z,\zeta),$$

where the function $H(z,\zeta)$ is defined by

$$(2.5)\qquad H(z,\zeta) := \begin{cases} z-\zeta & \text{for} \quad |\zeta| \le 1 \\ (z-\zeta)|\zeta|^{-1} & \text{for} \quad \infty > |\zeta| > 1 \\ 1 & \text{for} \quad \zeta = \infty \end{cases} .$$

This last function is called the <u>standard</u> <u>linear</u> <u>factor</u>. We note the $P \equiv 1$ is contained in every $\mathbf{P}_n$, $n \in \mathbf{N}$, and it is associated with the empty zero set $Z(P) = \phi$, while $P \equiv 0$ does not belong to any $\mathbf{P}_n$, $n \in \mathbf{N}$. There are many ways to normalize polynomials. Perhaps the most common is given by the monic polynomials $P(z) = z^n + \ldots \in \Pi_n$, $n \in \mathbf{N}$. However, the normalization (2.4) has the advantage that the polynomials $P \in \mathbf{P}_n$ remain bounded if some of their zeros tend to infinity, a property which does not hold in the case of monic polynomials. A polynomial $P \in \mathbf{P}_n$ is monic if and only if $Z(P) \subseteq \{|z| \le 1\}$.

For a finite set $A = A(n)$ of n points from $\overline{\mathbb{C}}$ there exists exactly one polynomial $P \in \mathbf{P}_n$ with $Z(P) = A|_{\mathbb{C}}$, where $A|_{\mathbb{C}}$ denotes the restriction of A to $\mathbb{C}$, which has already been defined above. This uniquely existing polynomial is denoted by $P(A;z)$.

We have seen in (2.2) and (2.3) that interpolation in Hermite's sense demands an appropriate order of differentiability at every point of the interpolation set A. By $F(A)$ we denote the collection of all functions f defined in a neighborhood of every point of $S(A)$ and being there at least $m-1$ times differentiable if the point in question is contained m-times $(m > 1)$ in A. If $a = \infty$, then we assume that $\tilde{f}(z) := f\left[\frac{1}{z}\right]$ is $m-1$ times differentiable at $z = 0$. All functions in $F(A)$ are assumed to have finite values and finite derivatives of appropriate order at every points of $S(A)$. The extension to the case of functions assuming the value infinity is postponed until Section 8.

With the notation introduced so far we can define <u>Hermite</u> <u>interpolation</u> (interpolation in Hermite's sense) in a unified and simple way for interpolation sets contained in $\mathbb{C}$. The extension to $\overline{\mathbb{C}}$ will be given below after the introduction of polynomials in the variable $\frac{1}{z-\zeta}$ with $\zeta \in \mathbb{C}$ fixed.

<u>Definition</u> 2.1' Let $A = A(n) := \{a_1, \ldots, a_n\}$, $n \in \mathbf{N}$, be an interpo-

lation set of n points $a_j \in \mathbb{C}$, $j = 1,\dots,n$, not necessarily all distinct. We say that a function $g \in F(A)$ <u>interpolates</u> a given function $f \in F(A)$ in the set A if

$$(2.6) \qquad f(z) - g(z) = P(A;z)h(z),$$

where h is supposed to be defined and bounded in a neighborhood of $S(A)$.

Statement (2.6) implies that the two functions f and g have a contact at every point $a \in S(A)$ of an order at least as large as the multiplicity of the point a in A. It is easy to see that (2.6) is equivalent to (2.2) and (2.3).

In order to extend Definition 2.1' to interpolation sets A containing infinity, we introduce for $\zeta \in \mathbb{C}$ and for sets $A = A(n) := \{a_1,\dots,a_n\}$, $a_j \in \overline{\mathbb{C}}$, polynomials in the variable $\frac{1}{z-\zeta}$ by

$$(2.7) \qquad P_\zeta(A;z) := \frac{P(A;z)}{H(z,\zeta)^n}$$

$$(2.8) \qquad \Pi_{n\zeta} := \left\{\frac{P(z)}{H(z,\zeta)^n} \; ; \quad P \in \Pi_n\right\}$$

$$(2.9) \qquad P_{n\zeta} := \left\{\frac{P(z)}{H(z,\zeta)^n} \; ; \quad P \in P_n\right\}$$

where $H(z,S)$ is the standard linear factor introduced in (2.5). We have $P_\infty(A;z) = P(A;z)$, $\Pi_{n\infty} = \Pi_n$, $P_{n\infty} = P_n$, $\Pi_{nS} = \left\{P\left[\frac{1}{z-\zeta}\right]; P \in \Pi_n\right\}$ if $\zeta \in \mathbb{C}$, and $P_\zeta(A;z) \in P_{n\zeta}$ for all $\zeta \in \overline{\mathbb{C}}$ and card $(A) \leq n$. In $P_{n\zeta}$, $\zeta \in \overline{\mathbb{C}}$, a polynomial P is uniquely determined by its zero set $Z(P)$. For $\zeta \in \overline{\mathbb{C}}$ and a given finite set A of points in $\overline{\mathbb{C}}$ the uniquely existing polynomial $P \in P_{n\zeta}$ having $A|_{\overline{\mathbb{C}}\setminus\{\zeta\}}$ as zero set is given by (2.7). Definition (2.7) implies a normalization of the polynomials in $P_{n\zeta}$, which is consistent with that introduced in (2.4) for the case $\zeta = \infty$.

Using the polynomials of $P_{n\zeta}$, $\zeta \in \overline{\mathbb{C}}$, on the right-hand side of (2.6) we can extend Definition 2.1' to interpolation sets containing infinity.

<u>DEFINITION</u> 2.1 Let $A = A(n) := \{a_1,\dots,a_n\}$, $n \in \mathbb{N}$, be an interpola-

tion set of n points $a_j \in \overline{\mathbb{C}}$, $j = 1,\ldots,n$, not necessarily all distinct. We say that a function $g \in F(A)$ <u>interpolates</u> a given function $f \in F(A)$ in the set A if

$$(2.10) \qquad f(z) - g(z) = P_\zeta(A;z)h(z),$$

where h is defined and bounded in a neighborhood of $S(A)$, and $\zeta \in \overline{\mathbb{C}}\backslash S(A)$.

<u>REMARK</u>: It is easy to see that the interpolation property - of the two functions f and g is independent of the actual choice of $\zeta \in \overline{\mathbb{C}}\backslash S(A)$.

Interpolation in a given set A defines an equivalence relation in $F(A)$. Two functions $f,g \in F(A)$ are considered equivalent if f interpolates g in A. It follows from the main result of the next section that if $S(A) \subset \mathbb{C}$ then every equivalence class in $F(A)$ can uniquely be represented by a polynomial P of degree card(A)-1.

3. Polynomial Interpolants

The well-known theorem on unique existence of polynomial interpolants is stated and proved. The proof is included since it is short and gives an opportunity to become accustomed to the new notation.

<u>THEOREM</u> 3.1 <u>For any interpolation set</u> $A = A(n+1) := \{a_1,\ldots,a_{n+1}\}$ <u>of</u> $n+1$ <u>points</u> $a_j \in \mathbb{C}$, <u>not necessarily all distinct</u>, <u>and a given function</u> $f \in F(A)$, <u>there exists exactly one polynomial</u> $P_n \in \Pi_n$ <u>that interpolates</u> f <u>in</u> A.

<u>DEFINITION</u> 3.2 The uniquely existing polynomial of Theorem 3.1 is called the <u>polynomial interpolant</u> to f in the set A, and it is denoted by $L_n(f,A;z)$.

<u>Proof of Theorem</u> 3.1: Let A contain exactly m distinct points $\{a_1,\ldots,a_m\}$ and each point a_j with a multiplicity m_j, $j = 1,\ldots,m$, i.e. $S(A) = \{a_1,\ldots,a_m\}$ and $\Sigma m_j = n+1$. Put $P_j(z) := P(A_j;z) \in \Pi_{n+1-m_j}$ with $A_j := A|_{\mathbb{C}\backslash\{a_j\}}$, $j = 1,\ldots,m$, and further $P_{j\ell}(z) := (z-a_j)^\ell P_j(z) \in \Pi_n$ for $j = 1,\ldots,m$, $\ell = 0,\ldots,m_j-1$. Each

matrix $\{P_{j\ell}^{(k)}(a_j)\}_{\ell k=0}^{m_j-1}$, $j = 1,\dots,m$, is upper triangular and all its diagonal elements are not equal to zero. Hence, these matrices are non-singular, and there exist constants $c_{\ell k}^{j}$, $j = 1,\dots,m$; $\ell, k = 0,\dots,m_j-1$, satisfying

$$(3.1) \qquad \sum_{i=0}^{m_j-1} c_{\ell i}^{j} P_{ji}^{(k)}(a_j) = \delta_{\ell k}$$

for $j = 1,\dots,m$; $\ell, k = 0,\dots,m_j-1$, where $\delta_{\ell k} = 1$ for $\ell = k$ and $\delta_{\ell k} = 0$ for $\ell \neq k$. With these constants $c_{\ell i}^{j}$ we define $n+1$ basic polynomials

$$(3.4) \qquad L_{j\ell}(z) := \sum_{i=0}^{m_j-1} c_{\ell i}^{j} P_{ji}(z) \in \Pi_n ,$$

$j = 1,\dots,m$; $\ell = 0,\dots,m_j-1$, which satisfy

$$(3.5) \qquad L_{j\ell}^{(k)}(a_i) = \delta_{ij}\delta_{\ell k} .$$

From (3.5) it follows that

$$(3.6) \qquad L(z) := \sum_{j=1}^{m} \sum_{\ell=0}^{m_j-1} f^{(\ell)}(a_j) L_{j\ell}(z) \in \Pi_n$$

interpolates f in the set A. This proves the existence of at least one polynomial interpolant.

In order to prove uniqueness let us assume that $L_j \in \Pi_n$, $j = 1,2$, interpolates f in A. Hence, there exist functions h_j, $j = 1,2$, defined and bounded in a neighborhood of $S(A)$ and $f - L_j = P(A_j;\cdot)h_j$, $j = 1,2$, which implies

$$(3.7) \qquad L_1(z) - L_2(z) = P(A;z)(h_1(z) - h_2(z)).$$

Since the left-hand side of this equation is a polynomial of degree not larger than n having $n + 1$ zeros, we have $L_1 \equiv L_2$. □

Let us now look at some special interpolation schemes:

(i) If in every $A(n)$, $n \in \mathbb{N}$, all points are distinct, then we have the case of Lagrange interpolation.

(ii) Let N be a sequence of points $\{a_1,\ldots,a_n,\ldots\}$ in $\mathbb{C}$, and $A(n)$ the first section of length n of this sequence, then we get a new interpolation set $A(n+1)$ by adding the point a_{n+1} to the last set $A(n)$. Schemes of this type define Newton interpolation and are called Newton interpolation schemes.

(iii) If all points in (2.1) are equal to a certain point $a_0 \in \mathbb{C}$, then the interpolation polynomials are the sections of the Taylor series expansion of the function f at a_0.

4. Cauchy-Jacobi Interpolants

Rational interpolants named after Cauchy and Jacobi will be defined in this section. They do not always exist, but if they exist then they are unique, (if we here neglect some minor problems that arise with the use of infinity as an interpolation point.) In order to allow infinity to be included as an ordinary interpolation point we introduce a notation for rational functions in the variable $\frac{1}{z-\zeta}$ with $\zeta \in \mathbb{C}$ fixed.

Having Theorem 3.1 in mind we may hope to get an analoguos theorem for rational interpolants. Since any rational function

$$(4.1) \qquad R(Z) = \frac{p_m z^m + \ldots + p_0}{q_n z^n + \ldots + q_0} \in \Pi_{mn}, \quad m,n \in \mathbb{N},$$

depends essentially on $m+n+1$ coefficients, the best we can expect is interpolation in $m+n+1$ arbitrary points $a_j \in \overline{\mathbb{C}}$. This expectation is the basis of the following definition.

DEFINITION 4.1 Given an interpolation set $A = A(n+m+1) := \{a_1,\ldots,a_{m+n+1}\}$ of $m+n+1$ points $a_j \in \overline{\mathbb{C}}$, not necessarily all distinct, $m, n \in \mathbb{N}$, and a function $f \in F(A)$, then a rational function $R \in \Pi_{mn}$ is called the Cauchy-Jacobi interpolant of f in A if R interpolates f in the set A.

One may also say $R \in \Pi_{mn}$ solves the <u>Cauchy-Jacobi</u> <u>interpolation</u> <u>problem</u> determined by f, A, and $m, n \in \mathbb{N}$. This formulation underscores the fact that the existence of a Cauchy-Jacobi interpolant is a problem in itself. Indeed, the interpolation problem is not always solvable. This is demonstrated by the following counterexample.

<u>EXAMPLE</u> 4.2 Let $A = A(3) := \{a_1, a_2, a_3\}$ be a set of three distinct point $a_j \in \mathbb{C}$, and let f be a function satisfying $f(a_1) = f(a_2) \neq f(a_3)$. Then the interpolation of f in the set A by functions of $\Pi_{1,1}$ is not possible. Indeed, all rational functions in $\Pi_{1,1}$ are Moebius transforms or constants. Since Moebius transforms are univalent mappings of $\overline{\mathbb{C}}$ on $\overline{\mathbb{C}}$, it is necessary that all three values $f(a_1)$, $f(a_2)$, and $f(a_3)$ are distinct or equal in order that they may be interpolated by functions of $\Pi_{1,1}$.

Although, we cannot have a simple, straight forward existence theorem for Cauchy-Jacobi interpolants, we can however prove uniqueness for arbitrary interpolation sets in $\mathbb{C}$. If infinity is among the interpolation points, then we have to impose the additional condition $m \geq n$.

<u>THEOREM</u> 4.3 <u>If two rational functions</u> $R_1, R_2 \in \Pi_{mn}$, $m, n \in \mathbb{N}$, <u>interpolate a function</u> $f \in F(A)$ <u>in a set</u> $A := \{a_1, \ldots, a_{m+n+1}\}$ <u>of</u> $m+n+1$ <u>points</u> $a_j \in \mathbb{C}$, <u>not necessarily all distinct, then</u> $R_1 \equiv R_2$. <u>If</u> $\infty \in A$, <u>then we have to assume</u> $m \geq n$ <u>in order to ensure</u> $R_1 \equiv R_2$.

The necessity of the condition $m \geq n$ in the last sentence of Theorem 4.3 is shown by

<u>EXAMPLE</u> 4.4 Let us consider $A = A(2) := \{0, \infty\}$ and $f(z) := 1/(1+z)$. Any rational function $a/(a+z) \in \Pi_{0,1}$, $a \in \mathbb{C}\setminus\{0\}$ will interpolate f in A. Hence, there is no uniqueness for $m = 0 < n = 1$.

<u>Proof of Theorem</u> 4.3: Let $R_j(z) = P_j(z)/Q_j(z)$, $P_j \in \Pi_m$, $Q_j \in P_n$, $j = 1, 2$, be two rational functions of Π_{mn} interpolating f in A, and let $\zeta \in \overline{\mathbb{C}}\setminus S(A)$. We have

$$(4.2) \qquad \frac{P_1(z)}{Q_1(z)} - \frac{P_2(z)}{Q_2(z)} = P_\zeta(A;z)\, h_1(z),$$

where h_1 is defined and bounded in a neighborhood of $S(A)$, and $P_\zeta(A;z)$ is the polynomial introduced in (2.7).

Let us first assume that $S(A) \subseteq \mathbb{C}$. Multiplying (4.2) by the two denominators Q_1 and $/Q_2$ yields

$$(4.3) \qquad (P_1Q_2 - P_2Q_1)(z) = P_\zeta(A;z)h_2(z),$$

where $h_2 := Q_1Q_2h_1$ is again bounded in a neighborhood of $S(A)$. On the left-hand side of (4.3) we have a polynomial of degree not greater than $m+n$ having $m+n+1$ zeros. Hence, this polynomial is identically zero, and we have $R_1 \equiv R_2$.

Let now $\infty \in A$, and let $k \in N$ be the multiplicity of ∞ in A. Further we assume $m \geq n$. Under these assumptions the function h_2 in (4.3) is bounded on $S(A) \cap \mathbb{C}$, and the right-hand side of (4.3) has a polar singularity of order not greater than $2n-k$ at infinity. Hence, the left-hand side of (4.3) is a polynomial of degree not greater than $2n-k$ having $m+n+1-k$ zeros. Since $m+n+1-k > 2n-k$, this polynomial has to be identically zero, which implies $R_1 \equiv R_2$. □

We have seen that interpolation at infinity by functions of Π_{mn} has special features if $m < n$. This is caused by the fact that the actual numerator and denominator degrees of the interpolant are determined by the behavior of f at infinity. In order to get rid of the special rule of infinity, we introduce rational functions in the variable $\frac{1}{z - \zeta}$ with $\zeta \in \mathbb{C}$ fixed. Their defintion is an extension of (2.7), (2.8), and (2.9) in the polynomial case. We denote

$$(4.4) \qquad \Pi_{mn\zeta} := \{P/Q;\ P \in \Pi_{m\zeta},\ Q \in P_{n\zeta}\},$$

$m,n \in N$, $\zeta \in \overline{\mathbb{C}}$. If $\zeta \in \mathbb{C}$, then a rational function $R \in \Pi_{mn\zeta}$ is of the form

$$(4.5) \qquad R(z) = \frac{p_m\left[\frac{1}{z-\zeta}\right]^m + \ldots + p_0}{q_n\left[\frac{1}{z-\zeta}\right]^n + \ldots + q_0}, \quad p_j, q_j \in \mathbb{C},$$

i.e. numerator and denominator are polynomials in $\frac{1}{z-\zeta}$. It is easy to see that $\Pi_{mn\infty} = \Pi_{mn}, \Pi_{mn\zeta} = \left\{R\left[\frac{1}{z-\zeta}\right]; R \in \Pi_{mn}\right\}$ for $\zeta \in \mathbb{C}$, $\Pi_{nn\zeta} = \Pi_{nn\eta}$ for all $\zeta, \eta \in \overline{\mathbb{C}}$ and $n \in \mathbb{N}$, but $\Pi_{mn\zeta} \neq \Pi_{mn\eta}$ if $m \neq n$ and $\zeta \neq \eta$. We have

$$(4.6) \qquad \Pi_{mn\zeta} \subseteq \Pi_{NN}, \quad N = \max(m,n),$$

for all $\zeta \in \overline{\mathbb{C}}$ and $m,n \in \mathbb{N}$.

The most often used values of ζ are $\zeta = \infty$ and $\zeta = 0$. In the first case we have the ordinary rational functions, in the later case rationals in $\frac{1}{z}$. If we simultaneously interpolate at zero and infinity we may, however, be forced to use a point $\zeta \in \mathbb{C}$ different from zero and infinity.

To be independent of the special rule of infinity we use the rational functions of $\Pi_{mn\zeta}$, $m,n \in \mathbb{N}$, instead of Π_{mn}. If we choose $\zeta \in \overline{\mathbb{C}}\backslash S(A)$, then all interpolation points in A can be considered as being of the same nature. Applying arguments as in the proof of Theorem 4.3 we can verify the following lemma.

<u>LEMMA</u> 4.5 <u>Let</u> $A = A(m+n+1) := \{a_1, \ldots, a_{m+n+1}\}$, $m,n \in \mathbb{N}$, <u>be an interpolation set of</u> $m+n+1$ <u>points</u> $a_j \in \overline{\mathbb{C}}$, <u>not necessarily all distinct</u>, <u>and</u> $\zeta \in \overline{\mathbb{C}}\backslash S(A)$ <u>a fixed point</u>. <u>If both function</u> R_1, $R_2 \in \Pi_{mn\zeta}$ <u>interpolate a function</u> $f \in F(A)$ <u>in</u> A, <u>then</u> $R_1 \equiv R_2$.

<u>REMARK</u> 1) It follows from Example 4.4 that if $\zeta \in S(A)$, then the uniqueness of a rational function $R \in \Pi_{mn\zeta}$ interpolating a function $f \in F(A)$ is only ensured if $m \geq n$.

2) Since $\Pi_{nn\zeta} = \Pi_{nn}$ for all $\zeta \in \overline{\mathbb{C}}$, it follows from Theorem 4.3 that in case of $m = n$ a Cauchy-Jacobi interpolant is always unique if it exists. In this case it is also independent of ζ. That this last sentence is not true in case of $m \neq n$ is demonstrated by

Example 4.6 Let $A = A(2) := \{0,0\}$ and $f(z) := 1+z$. We give the interpolants to f in A from the four sets of rational functions $\Pi_{1,0,\zeta=1}$, $\Pi_{1,0,\zeta=-1}$, $\Pi_{0,1,\zeta=1}$, and $\Pi_{0,1,\zeta=-1}$. They are

$$(4.7a) \qquad R_1(z) = \frac{-1}{z-1},$$

$$(4.7b) \qquad R_2(z) = 2 - \frac{1}{z+1},$$

$$(4.7c) \qquad R_3(z) = \frac{1}{2 + 1/(z-1)} = \frac{1-z}{1-2z},$$

$$(4.7d) \qquad R_4(z) = \frac{1}{1/(z+1)} = 1 + z,$$

respectively, and indeed all different.

We close the section with a historic remark. It seems that for interpolation sets with only distinct points in $\mathbb{C}$ the first explicit formula for interpolating rational functions has been published by CAUCHY (1821). JACOBI (1846) simplified Cauchy's formula and included the case of confluent interpolation points as well as the case of rational functions in the variable $\frac{1}{z-\zeta}$, with $\zeta \in \mathbb{C}$ fixed. Both authors have apparently been unaware of the existence problem for rational interpolants. Actually, they solved the linearized rational interpolation problem, which is now more commonly known as multipoint Padé approximation, and which will be defined and discussed in the next section. As is shown in Lemma 5.3 below, these functions are identical to the interpolants of Definition 4.1 if the latter exist. The first rigorous treatment of the existence problem for rational interpolants is contained in KRONECKER (1881), where we also find algorithms for the calculation of the interpolants, which are based on continued fractions expansion.

5. Multipoint Padé Approximants

Multipoint Padé approximants, which are also known as linearized or modified rational interpolants, are defined and their unique existence is shown.

The lack of a general existence theorem for Cauchy-Jacobi interpolants makes it worth while to study this modified definition of rational interpolants. The definition is based on the next lemma. There we consider not only rational functions and polynomials in the variable z, but also in the variable $\frac{1}{z-\zeta}$ with $\zeta \in \mathbb{C}$ fixed, using the notation introduced in (2.7) - (2.9), and (4.5).

LEMMA 5.1 *Let* $A = A(m+n+1) := \{a_1, \ldots a_{m+n+1}\}$, $m,n \in \mathbb{N}$, *be a set of* $m+n+1$ *points* $a_j \in \overline{\mathbb{C}}$, *not necessarily all distinct*, $\zeta \in \overline{\mathbb{C}}\setminus S(A)$ *and* $f \in F(A)$. *There exist two polynomials* $P_{mn} \in \Pi_{m\zeta}$, *and* $Q_{mn} \in P_{n\zeta}$ (i.e. P_{mn} *and* Q_{mn} *are polynomials in* $\frac{1}{z-\zeta}$ *if* $\zeta \neq \infty$) *such that*

$$Q_{mn}(z)f(z) - P_{mn}(z) = P_\zeta(A;z)h(z), \tag{5.1}$$

where h *is defined and bounded in the neighborhood of* $S(A)$. *The rational function*

$$R_{mn}(f,A,\zeta;z) := \frac{P_{mn}(z)}{Q_{mn}(z)} \tag{5.2}$$

is uniquely determined by (5.1). *If* $m = n$, *then* $R_{nn}(f,A,\zeta;z)$ *is independent of the point* $\zeta \in \overline{\mathbb{C}}\setminus S(A)$.

Proof: We shall first prove the existence of polynomials P_{mn} and Q_{mn}. Let us assume $\zeta \neq \infty$. The case $\zeta = \infty$ will then follow. By A_ζ we denote the image of A under the mapping $z \to \frac{1}{z-\zeta}$. Hence, A_ζ is an interpolation set of $m+n+1$ points and $S(A_\zeta) \subseteq \mathbb{C}$. Denote by

$$P_\ell(z) := \sum_{j=1}^{m+n} b_{\ell j} z^j \quad , \ \ell = 0,\ldots,n \tag{5.3}$$

the uniquely existing polynomials interpolating the function $z^\ell f\left[\zeta + \frac{1}{z}\right]$ in the set A_ζ. For any choice of coefficients $c_\ell \in \mathbb{C}$, $\ell = 0,\ldots,n$, and $Q(z) := \sum_{\ell=0}^n c_\ell z^\ell \in \Pi_n$, the polynomial

$$P(z) := \sum_{\ell=0}^{n} c_\ell P_\ell(z) = \sum_{j=0}^{m+n} \sum_{\ell=0}^{n} c_\ell b_{\ell j} z^j \tag{5.4}$$

interpolates $Q(z)f\left[\zeta + \frac{1}{z}\right]$ in A_ζ. If we choose the coefficients

c_ℓ, $\ell = 0,\dots,n$, such that

$$(5.5) \qquad \sum_{\ell=0}^{n} c_\ell b_{\ell j} = 0 \quad \text{for} \quad j = m+1,\dots,m+n,$$

which is always possible in a non-trivial way (i.e. not all $c_\ell = 0$), then $P \in \Pi_m$ and $Q \not\equiv 0$. Substituting the variable z by $\frac{1}{z-\zeta}$ we have $P_{mn}(z) := P\left[\frac{1}{z-\zeta}\right] \in \Pi_{m\zeta}$ and $Q_{mn}(z) := Q\left[\frac{1}{z-\zeta}\right] \in \Pi_{n\zeta}$. Both polynomials satisfy (5.1), and if we multiply both with an appropriate constant we have also $Q_{mn} \in P_{n\zeta}$. This completes the proof of existence for $\zeta \in \mathbb{C}$. The case $\zeta = \infty$ is covered if we take $A = A_\zeta$.

The proof of uniqueness of the function (5.2) is practically the same as the first part of the proof of Theorem 4.3.

It remains to verify that in case of $m = n$ the function $R_{nn}(f,A,\zeta;z)$ is indeed independent of $\zeta \in \overline{\mathbb{C}}\setminus S(A)$. Let $\zeta,\eta \in \overline{\mathbb{C}}\setminus S(A)$ and

$$(5.6) \qquad R_{nn}(f,A,\eta;z) = \frac{P(z)}{Q(z)}, \; P \in \Pi_{n\eta}, \; Q \in P_{n\eta}.$$

From the definition of $\Pi_{n\zeta}$ and $P_{n\zeta}$ in (2.8) and (2.9) it follows that $P^*(z) := \left[\frac{H(z,\eta)}{H(z,\zeta)}\right]^n P(z) \in \Pi_{n\zeta}$ and $Q^*(z) := \left[\frac{H(z,\eta)}{H(z,\zeta)}\right]^n Q(z) \in P_{n\zeta}$. Hence, the polynomials P^* and Q^* satisfy (5.1), and because of the uniqueness proved in the first part we have

$$(5.7) \qquad R_{nn}(f,A,\eta;z) = \frac{P(z)}{Q(z)} = \frac{P^*(z)}{Q^*(z)} = R_{nn}(f,A,\zeta;z). \qquad \square$$

<u>DEFINITION</u> 5.2 Let $A = A(m+n+1) := \{a_1,\dots,a_{m+n+1}\}$, $m,n \in \mathbb{N}$, be a set of $m+n+1$ points $a_j \in \overline{\mathbb{C}}$, not necessarily all distinct, $\zeta \in \overline{\mathbb{C}}\setminus S(A)$, and $f \in F(A)$. The uniquely existing rational function $R_{mn}(z) = R_{mn}(f,A,\zeta;z) \in \Pi_{mn\zeta}$ of Lemma 5.1 is called the <u>multipoint Padé approximant</u> (developed in powers of z if $\zeta = \infty$, and in powers of $\frac{1}{z-\zeta}$ if $\zeta \neq \infty$).

<u>REMARKS</u>: 1) While the polynomial P_{mn} interpolates $Q_{mn}f$ in the set A, the multipoint Padé approximant R_{mn} may not interpolate f

in every point of A. These possible interpolation defects are the reason why R_{mn} is called an approximant and not an interpolant. We shall have a closer look at interpolation defects in the next section.

2) In contrast to the rational function (5.2) the two polynomials Q_{mn} and P_{mn} are in general not unique. It follows from (5.5) and the normalization of polynomials in $P_{n\zeta}$ that they are unique if and only if $(b_{\ell,j+m})_{\ell=0,\ldots,n;\, j=1,\ldots,n}$ is of rank n.

3) Lemma 5.1 shows that the case $m = n$ is special in the sense that all points of $\overline{\mathbb{C}}$ are equal with respect to interpolation, and the approximant is independent of the point $\zeta \in \overline{\mathbb{C}}\setminus S(A)$. This independence does not hold if $m \neq n$, as we can easily see from Example 4.6. The rational interpolants given there in (4.7) are multipoint Padé approximants, which can easily be verified, but it also follows from the next lemma. This lemma shows that multipoint Padé approximants are the best we can try in order to find a solution of a Cauchy-Jacobi interpolation problem.

LEMMA 5.3 *If for an interpolation set* $A = A(m+n+1)$ *of* $m+n+1$ *points in* $\overline{\mathbb{C}}$, *not necessarily distinct*, $m,n \in \mathbb{N}$, $\zeta \in \overline{\mathbb{C}}\setminus S(A)$, *and a function* $f \in F(A)$, *the Cauchy-Jacobi interpolation problem is solvable in* $\Pi_{mn\zeta}$, *then the multipoint Padé approximant* $R_{mn}(f,A,\zeta;z)$ *is the solution.*

Proof: Let $R = P/Q \in \Pi_{mn\zeta}$, $P \in \Pi_{m\zeta}$, $Q \in P_{n\zeta}$, interpolate f in A. Then p interpolates Qf in A. Hence, P and Q satisfy (5.1) and $R \equiv R_{mn}(f,A,\zeta;.)$. □

Of course, if the Cauchy-Jacobi problem is not solvable, the multipoint Padé approximant cannot interpolate in all points of A. To illustrate this situation, we go back to Example 4.2. Putting $Q_{1,1}(z) := H(z,a_3)$ and $P_{1,1}(z) := f(a_1)H(z,a_3)$, it follows that

$$(5.8)\qquad Q_{1,1}(a_j)f(a_j) - P_{1,1}(a_j) = 0\ ,\quad j = 1,2,3.$$

Hence, the 1,1-multipoint Padé approximant in this example is

$$(5.9)\qquad R_{1,1}(z) = \frac{P_{1,1}(z)}{Q_{1,1}(z)} \equiv f(a_1).$$

which interpolates in a_1 and a_2, but not in a_3.

Definition 5.2 is rather general. Let us therefore look at the two most common special cases:

(i) If the interpolation set $A = A(m+n+1) := \{0,\ldots,0\}$, $m,n \in \mathbb{N}$, consists of $m+n+1$ repetitions of the point $z = 0$, and $f(z)$ is given as a power series $\Sigma a_j z^j$, then Definition 5.2 gives us the <u>ordinary</u> m,n <u>Padé</u> <u>approximant</u>, which is generally denoted by $[m/n](z)$ (cf. PERRON (1929)). We have $[m/n] = P/Q$, $P \in \Pi_m$, $Q \in P_n$, and

$$(5.10) \qquad (Qf-P)(z) = O(z^{m+n+1}) \quad \text{for} \quad z \to 0.$$

(ii) If $A = A(m+n+1) := \{\infty,\ldots,\infty\}$, $m,n \in \mathbb{N}$, consists of $m+n+1$ repetitions of the point $z = \infty$, $\zeta = 0$, then Definition 5.2 gives the ordinary m,n Padé approximant developed at infinity in powers of $\frac{1}{z}$. If $f(z) = \Sigma a_j z^{-j}$, then we have $[m/n](z) = \dfrac{P\left[\frac{1}{z}\right]}{Q\left[\frac{1}{z}\right]}$, $P \in \Pi_m$, $Q \in P_n$, and

$$(5.11) \qquad Q\left[\frac{1}{z}\right]f(z) - P\left[\frac{1}{z}\right] = Q(z^{-(m+n+1)}) \quad \text{for} \quad z \to \infty .$$

Both cases considered here are immediate generalizations of sections of the Taylor series of $f(z)$ at $z = 0$ or $z = \infty$ to the field of rational functions.

6. <u>Interpolation</u> <u>Defects</u>

Interpolation defects, and also normality defects, are defined, and sufficient conditions are given which ensure that a multipoint Padé approximant solves a given Cauchy-Jacobi interpolation problem.

While the general existence theorem is the great advantage of multipoint Padé approximants, the possibility of interpolation defects is their main drawback. As it has been shown by Example 4.2, interpolation defects are in general not avoidable. The only thing we can obtain are criteria that tell us whether interpolation defects are possible, and if so, how large they can be.

Let again $A = A(m+n+1) := \{a_1,\ldots,a_{m+n+1}\}$, $m,n \in \mathbb{N}$, be an interpolation set of $m+n+1$ points in $\overline{\mathbb{C}}$ not necessarily all distinct, and $\zeta \in \overline{\mathbb{C}}\setminus S(A)$. By $\Pi_\zeta(A)$ we denote the set $\{P; P \in \Pi_{k\zeta}, k \in \mathbb{N}, P(z) \neq 0 \text{ for } z \in \overline{\mathbb{C}}\setminus S(A)\}$ of polynomials in the variable $\frac{1}{z-\zeta}$ or z. We consider $f \in F(A)$ and the multipoint Padé approximant $R_{mn}(z) = R_{mn}(f, A, \zeta, z)$ to f. Let $P_0(z) = P_0(f, R_{mn}, A, \zeta; z) \in \Pi_\zeta(A)$ be the polynomial of largest degree in $\Pi_\zeta(A)$ satisfying

(6.1) $$(f - R_{mn})(z) = P_0(z)h(z),$$

where h is supposed to be bounded and defined on a neighborhood of $S(A)$. It is easy to see that P_0 is unique and $\deg(P_0) < \infty$ if f is analytic on $S(A)$ and not identical to R_{mn}. The order of a zero of P_0 at $a \in S(A)$ is the order of contact between f and R_{mn} at the point a, and $\deg(P_0)$ gives the total order of contact between f and R_{mn} on $S(A)$.

<u>DEFINITION</u> 6.1 Let A, ζ, $m,n \in \mathbb{N}$, f, and R_{mn} be defined as above. For $a \in S(A)$ the number

(6.2a) $$d(a) := \text{card } [A|_{\{a\}}\setminus Z(P_0)]$$

is called the <u>order of interpolation defect</u> of the multipoint Padé approximant R_{mn} to f at a, the number

(6.2b) $$d := \text{card } [A\setminus Z(P_0)]$$

is called <u>total order of interpolation defect</u>, and

(6.3) $$d_0 := \text{card } [(A\setminus Z(P_0)) \cup (Z(P_0)\setminus A)]$$

is called the <u>order of normality defect</u>. The Padé approximant R_{mn} is called <u>normal</u> if $d_0 = 0$.

The Padé approximant R_{mn} interpolates f in A if and only if $d = 0$, i.e. if at every point $a \in S(A)$ the order of contact of f and R_{mn} is at least as large as the multiplicity of a in A. If

R_{mn} is normal, then at every point $a \in S(A)$ the order of contact between f and R_{mn} is exactly the same as the multiplicity of a in A.

The following lemma is often helpful in estimating interpolation defects.

LEMMA 6.2 *The multipoint Padé approximant* $R_{mn}(z) = R_{mn}(f,A,\zeta,z)$ *can have an interpolation defect of order* $\ell > 0$ *at a point* $a \in A$ *only if every denominator polynomial* Q_{mn} *of* R_{mn} *has a zero of order at least* ℓ *at* a. (*Note that the polynomial* Q_{mn} *in* (5.2) *may not be unique*.)

Proof: Let $a \in S(A)$ and let $k > 0$ be the multiplicity of a in A. We assume without loss of generality that $a \neq \infty$ and $\zeta = \infty$. The cases of $a = \infty$ and/or $\zeta \neq 0$ can be treated in exactly the same way by using polynomials from $\Pi_{n\zeta}$ instead of polynomials from Π_n.

Let $Q_{mn} \in \Pi_n$ have a zero of order $\ell \geq 0$ at a, and put $Q_{mn}(z) = (z-a)^{\ell}Q(z)$, $Q(a) \neq 0$, $Q \in \Pi_{n-\ell}$. If $\ell \geq k$ nothing has to be proved, let us therefore assume $\ell < k$. From (5.1) it follows that $P_{mn} \in \Pi_m$ must also have a zero of order at least ℓ at a. Hence, $P_{mn}(z) = (z-a)^{\ell}P$, $P \in \Pi_{m-\ell}$, and from (5.1) follows

$$\left[f - \frac{P}{Q}\right](z) = \frac{P(A;z)}{(z-a)^{\ell}}\,\frac{h(z)}{Q(z)}, \tag{6.4}$$

where $P(A;z)(z-a)^{-\ell}$ has a zero of order $k-\ell$ at the point a and h/Q is bounded in a neighborhood of a. This implies that f and $R_{mn} = P/Q$ have a contact of order at least $k-\ell$ at a, and therefore the interpolation defect at a cannot be greater than ℓ. □

From Lemma 6.2 we get as a corollary a sufficient condition for solvability of the Cauchy-Jacobi interpolation problem.

COROLLARY 6.3 *If* $Q_{mn}(a) \neq 0$ *for every point* $a \in A$, *then the multipoint Padé approximant* $R_{mn} = \frac{P_{mn}}{Q_{mn}} = R_{mn}(f,A,\zeta;.)$ *interpolates* f *in* $A = A(m+n+1)$.

We remark that with respect to normality the situation is somewhat different for ordinary and multipoint Padé approximants. In the latter case there does in general not exist a block structure. For Newton interpolation schemes, however, there exists a block structure, which has been investigated in detail by CLAESSENS (1976). The existence of interpolation defects or non-normalities is for fixed $m,n \in \mathbb{N}$ independent of the uniquness of the two polynomials $P_{mn} \in \Pi_{m\zeta}$ and $Q_{mn} \in P_{n\zeta}$ in (5.2). This is shown by the next two examples. In the first one the polynomials $P_{1,1}$ and $Q_{1,1}$ are unique, while there exists an interpolation defect. In the second example the situation is reversed.

Example 6.4 Let $A = A(z) := \{-1,0,1\}$ and $f(z) := z^2$, then from

$$(6.5) \qquad [(q_0+q_1z)z^2 - (p_0+p_1z)]_{z=-1,0,1} = 0$$

it follows that

$$(6.6) \qquad \begin{aligned} q_0 + q_1 &= p_0 + p_1 \\ q_0 - q_1 &= p_0 - p_1 \\ 0 &= p_0 \ , \end{aligned}$$

which implies $p_0 = q_0 = 0$ and $q_1 = p_2$. From the condition $Q_{1,1}(z) = q_0 + q_1z \in P_1$ we finally get $q_1 = 1$. Hence, the polynomials $P_{1,1}(z) = z$ and $Q_{1,1}(z) = z$ are uniquely determined, but the multipoint Padé approximant

$$(6.7) \qquad R_{1,1}(z) = \frac{P_{1,1}(z)}{Q_{1,1}(z)} = 1$$

has an interpolation defect of order 1 at $a = 0$.

Example 6.5 Let again $A = A(3) := \{-1,0,1\}$, but now we consider $f(z) := 1+z-z^3$. Any pair of identical polynomials $P_{1,1}(z) := Q_{1,1}(z) := c_0+c_1z$, $c_0,c_1 \in \mathbb{C}$, satisfies

$$(6.8) \qquad [Q_{1,1}(z)(1+z-z^3) - P_{1,1}(z)]_{z=-1,0,1} = 0,$$

Hence, the polynomials $P_{1,1} \in \Pi$, $Q_{1,1} \in P_1$ are not unique. However, the multipoint Padé approximant $R_{1,1}(z) \equiv 1$ perfectly interpolates f in A, moreover, $R_{1,1}$ is normal with respect to f and A.

We note that the definition of normality given in Definition 6.1 is somewhat broader than the usual one, which takes in consideration the whole Padé table (See for instance PERRON (1929).) Our definition implies unlike the classical one that the elements of the antidiagonal of a block in the Padé table of ordinary Padé approximants have to be considered normal, however all the other elements of a block are non-normal.

7. Rational Interpolants with Prescribed Poles

Rational interpolants with prescribed poles are defined and their unique existence is proved.

In the case of multipoint Padé approximants both numerator and denominator polynomial are chosen with the aim to have a contact as high as possible with the function f to be interpolated. While this gives us powerful interpolants, it may at the same time lead to insurmountable difficulties, if one tries to prove the convergence of sequences of such interpolants or approximants. The reason for these difficulties lies in the rather complicated connections between the function f, the interpolation set A, and the location of the zeros of the denominator polynomial Q_{mn}. Knowledge about the latter is essential for proofs of convergence.

Since in the case of Padé approximants and Cauchy-Jacobi interpolants the location of the poles depends on the specific function f to be interpolated, we call them rational interpolants or approximants with <u>free poles</u>. The poles are free to adapt to the function f. In contrast to this type of rational functions we will now consider functions with <u>prescribed</u> poles.

It will turn out that methodologically rational interpolants with prescribed poles are closer related to polynomial interpolants than to rational interpolants with free poles.

<u>LEMMA</u> 7.1 <u>Let</u> $A = A(n+1) := \{a_1, \ldots, a_{n+1}\}$ <u>and</u> $B = B(n) := \{b_1, \ldots b_n\}$, $n \in N$, <u>be two disjoint sets of</u> $n+1$ <u>and</u> n <u>points</u> a_j,

$b_j \in \overline{\mathbb{C}}$, <u>and let</u> $f \in F(A)$. <u>There uniquely exists a rational function</u> $R_n(z) = R_n(f,A,B;z) \in \Pi_{nn}$ <u>interpolating</u> f <u>in the set</u> A <u>and having all its poles in</u> B. <u>The multiplicity of a pole of</u> R_n <u>is not allowed to exceed the multiplicity of the corresponding point in</u> B.

<u>Proof</u>: Let us first assume that $S(A) \cup S(B) \subset \mathbb{C}$. Then

$$(7.1) \qquad R_n(z) := \frac{L_n(Qf,A;z)}{Q(z)} \in \Pi_{nn}, \quad Q := P(B;.),$$

where $L_n(Qf,A;z) \in \Pi_n$ is the n-degree interpolation polynomial interpolating Qf in A (see Definition 3.2). Hence R_n interpolates f in A.

If $\zeta \notin S(A) \cup S(B)$, then let A_ζ and B_ζ be the images of A and B under the mapping $z \to w = \frac{1}{z-\zeta}$ and

$$(7.2) \qquad R_\zeta(w) = \frac{L_n(Qf,A_\zeta;w)}{P(B_\zeta;w)}, \quad Q := P(B_\zeta;.).$$

We note that $A_\zeta \subset \mathbb{C}$ and $B_\zeta \subset \mathbb{C}$. It is easy to see that

$$(7.3) \qquad R_n(z) := R_\zeta\left[\frac{1}{z} + \zeta\right] \subset \Pi_{nn}$$

interpolates f in A. This establishes the existence of $R_n(f,A,B;z) \in \Pi_{nn}$ in the general case.

In order to prove uniqueness we assume that two rational functions $R_j = P_j/Q_j$, $P_j \in \Pi_n$, $Q_j \in P_n$, $j = 1,2$, interpolate f in A and have all their poles in B. Hence, we have

$$(7.4) \qquad \frac{P_1(z)}{Q_1(z)} - \frac{P_2(z)}{Q_2(z)} = P_\zeta(A;z)h(z),$$

where $\zeta \in \overline{\mathbb{C}}\setminus(S(A) \cup S(B))$ and h is defined and bounded in a neighborhood of $S(A)$. Let $k \in N$ be the multiplicity of ∞ in B ($k = 0$, if $\infty \notin B$) and $n_j := \deg(Q_j)$, $j = 1,2$. We have $\mathrm{card}(B|_{\mathbb{C}}) = n-k$, and

(7.5) $\quad H_j(z) := P(B;z)/Q_j(z) \in P_{n-k-n_j}, \; j = 1,2.$

Since R_j can have only a pole of order k at $z = \infty$, we have

(7.6) $\quad \deg(P_j) \le n_j + k, \; j = 1,2.$

Multiplying (7.4) by $P(B;z)$ we get

(7.7) $\quad (P_1H_1 - P_2H_2)(z) = P_\zeta(a;z)P(B;z)h(z).$

Let us first assume $\infty \notin A$. From (7.5) and (7.6) it follows that deg $(H_jP_j) \le n$, $j = 1,2$. Hence, the left-hand side of (7.7) is a polynomial of degree not greater than n having $n+1$ zeros. This implies $P_1H_1 \equiv P_2H_2$ and thereby $R_1 \equiv R_2$.

Let us now assume $\infty \in A$ and let $\ell \in \mathbb{N}$ be the multiplicity of ∞ in A. Since h is defined and bounded near infinity the right-hand side of (7.7) has a pole at ∞ of order at most $n-\ell$, and therefore

(7.8) $\quad \deg(P_1H_1 - P_2H_2) \le n-\ell.$

But since card $(A|_{\mathbb{C}}) = n+1-\ell$, the polynomial $P_1H_1 - P_2H_2$ has $n+1-\ell$ zeros, which implies $P_1H_1 \equiv P_2H_2$ and $R_1 \equiv R_2$. □

<u>DEFINITION</u> 7.2 Let $A = A(n+1) := \{a_1,\ldots,a_{n+1}\}$ and $B = B(n) := \{b_1,\ldots,b_n\}$ be two disjoint sets of $n+1$ and n points $a_j, b_j \in \overline{\mathbb{C}}$, $n \in \mathbb{N}$, and $f \in F(A)$. The uniquely existing function $R_n(z) := R_n(f,A,B;z) \in \Pi_{nn}$ interpolating f in A and having its poles in B is called the <u>rational interpolant of</u> f <u>interpolating in</u> A <u>with prescribed poles in</u> B.

<u>REMARK</u>: In Definition 7.2 we assume that the cardinality of the two sets A and B differs exactly by one. This leads normally to a rational function in Π_{nn} with numerator and denominator degree equal to n. In order to get different degrees one has to prescribe $|m-n|$ poles or zeros at infinity. They have to be poles if $m > n$ and zeros if $n > m$.

8. Rational Interpolation of the Value Infinity

In this last section the definitions of Section 4, 5, and 7 are extended to interpolation in polar singularities.

Given a point $a \in \overline{\mathbb{C}}$, let the function f be defined and bounded in a punctured neighborhood of a. At the point a itself f may be infinity. We say that f is of polar order $d = d(f,a) \geq 1$ at a $(a \neq \infty)$ if $(z-a)^d f(z)$ is finite at a, but $(z-a)^{d-1} f(z)$ is not. In case of $a = \infty$ we have to consider the two functions $z^{-d} f(z)$ and $z^{-d+1} f(z)$. We say that f is of polar order $d(f,a) = 0$ at $a \in \overline{\mathbb{C}}$ if f is defined and bounded in a full neighborhood of a.

Let $A = A(n)$ be an interpolation set of n points in $\overline{\mathbb{C}}$. A function f is called polar on $S(A)$ if it is of finite polar order at every point $a \in S(A)$. The total polar order on $S(A)$ is defined as

$$(8.1) \qquad d(f,A) := \sum_{a \in S(A)} d(f,a).$$

Note that our definition of a function to be polar at a point is somewhat more general than the usual definition of a pole. However, in case of a meromorphic function f both definitions coincide.

By $D = D(f,A)$ we denote the finite set in which every point $a \in S(A)$ is exactly $d(f,a)$ times repeated. Thus, $\operatorname{card}(D) = d(f,A)$.

DEFINITION 8.1 Let $A = A(n) := \{a_1,\dots,a_n\}$ be an interpolation set of n points $a_j \in \overline{\mathbb{C}}$, not necessarily all distinct. By $M(A)$ we denote the set of all functions f polar on $S(A)$, and for $\zeta \in \overline{\mathbb{C}} \setminus S(A)$, $f \in M(A)$, and $D = D(f,A)$, we have

$$(8.2) \qquad P_\zeta(D;z) f(z) \in F(A).$$

It is easy to see that the definition of $M(A)$ does not depend on $\zeta \in \overline{\mathbb{C}} \setminus S(A)$

From (8.2) it follows that if $a \in S(A)$, $a \neq \infty$, is m times contained in A and $d = d(f,a) > 0$, then there exist m coefficients $f_j \in \mathbb{C}$, $j = -d, 1-d, \dots, m-1-d$, such that

$$(8.3) \qquad f(z) = \sum_{j=-d}^{m-1-d} f_j(z-a)^j + o(|z-a|^{m-1-d}),$$

where $o(\ldots)$ means that this term goes faster to zero than the expression in the brackets.

In the present section the set $M(A)$ takes the place of $F(A)$ in the preceding sections. From (8.2) it is apparent that the assumptions made in Definition 8.1 ensure that all informations necessary for interpolation in the set A is indeed available for functions $f \in M(A)$. We note that every function meromorphic on $S(A)$ belongs to $M(A)$ in the same way as earlier every function analytic on $S(A)$ belongs to $F(A)$. Using functions from $M(A)$ we get the following extension of Definition 2.1.

<u>DEFINITION</u> 8.2 Let the interpolation set A be as in Definition 8.1. We say that the function $g \in M(A)$ <u>interpolates</u> a given function $f \in M(A)$ in the set A if $\tilde{g}(z) := P_\zeta(D;z)g(z) \in F(A)$ interpolates $\tilde{f}(z) := P_\zeta(D;z)f(z) \in F(A)$ in the set A, where $D := D(f,A)$ and $\zeta \in \overline{\mathbb{C}} \setminus S(A)$.

Basically in Definition 8.2 interpolation of functions $f \in M(A)$ has been put down to interpolation of functions in $F(A)$ by premultiplying with the polynomial $P_\zeta(D;z)$. The expansion (8.3) shows that by doing so we have indeed the appropriate order of contact between the two functions f and g at every point $a \in S(A)$.

We shall now consider extensions of the definitions of Cauchy-Jacobi interpolants, multipoint Padé approximants, and rational interpolants with prescribed poles. Using Definition 8.2 we have only to substitute $F(A)$ by $M(A)$ in Definition 4.1, 5.2, and 7.2, which we assume for the remainder of the section.

Let us first look at Cauchy-Jacobi interpolants. It is easy to see that they only exist if the denominator degree $n \in \mathbf{N}$ is at least as large as the total polar order $d(f,A)$ of the function f to be interpolated. The situation with respect to sufficient conditions for existence remains exactly the same as before. We summerize these facts in the following theorem.

<u>THEOREM</u> 8.3 <u>Let</u> $A = A(k) := \{a_1,\ldots,a_k\}$ <u>be an interpolation set of</u> k <u>points</u> $a_j \in \overline{\mathbb{C}}$, <u>and</u> $f \in M(A)$ <u>a function with total polar order</u> $d = d(f,A)$.

(i) For any rational function $R \in \Pi_{mn\zeta}$, $m,n \in \mathbb{N}, \zeta \in \overline{\mathbb{C}} \backslash S(A)$: interpolating f in the set A it is necessary that $n \geq d$.

(ii) If $m+n = d+k-1$ and $R \in \Pi_{mn\zeta}$, $\zeta \in \overline{\mathbb{C}} \backslash S(A)$, interpolates f in the set A, then R is unique.

(iii) Let $m+n = d+k-1$ and $\zeta \in \overline{\mathbb{C}} \backslash S(A)$. A rational function $R \in \Pi_{mn\zeta}$ interpolating f in the set A exists if and only if a rational function $\tilde{R} \in \Pi_{m,n-d,\zeta}$ exists which interpolates $\tilde{f}(z) := P_\zeta(D;z)f(z) \in F(A)$ in the set A, where $D := D(f,A)$.

Proof: Part (i) follows immediately from Definition 8.2, and part (ii) follows from Lemma 4.5 together with Definition 8.2. Part (iii) can be proved in the same way as Lemma 4.5 if we remember that the denominator of every rational function interpolating f in A contains the polynomial $P_\zeta(D;z)$ as factor. □

If we do not restrict ourselves to rational functions $R \in \Pi_{mn\zeta}$ with $\zeta \in \overline{\mathbb{C}} \backslash S(A)$, as we did in Theorem 8.3, then in order to have uniquenss of the Cauchy-Jacobi interpolant we have to assume the additional condition $m \geq n-d$, which substitutes the condition $m \geq n$ in Theorem 4.3.

In case of multipoint Padé approximants we find a situation that is totally analogous. We summarize that results in the following lemma, which is essentially a corollary to Lemma 5.1, and therefore will not be proved here.

LEMMA 8.4 Let $A = A(k)$, $k \in \mathbb{N}$, $f \in M(A)$, and $d = d(f,A)$ be defined as in Theorem 8.3, $\zeta \in \overline{\mathbb{C}} \backslash S(A)$. If $m+n = d+k-1$, $m,n \in \mathbb{N}$, then there uniquely exists a rational function

$$(8.4) \qquad R_{mn}(z) = R_{mn}(f,A,\zeta;z) = \frac{P_{mn}(z)}{Q_{mn}(z)} \in \Pi_{mn\zeta}$$

such that the polynomials $P_{mn} \in \Pi_{m\zeta}$, $Q_{mn} \in P_{n\zeta}$ satisfy

$$(8.5) \qquad Q_{mn}(z)f(z)-P_{mn}(z) = P_\zeta(A;z)h(z),$$

where h *is defined and bounded in a neighborhood of* $S(A)$. *If* $m = n$, *then* $R_{nn}(f,A,\zeta;z)$ *is independent of the point* $\zeta \in \overline{\mathbb{C}}\backslash S(A)$.

REMARKS: 1) It follows from (8.5) that $P_\zeta(D;z)$, $D := D(f,A)$, is a factor of Q_{mn}.

2) In the same way as in Lemma 5.3 it can be shown that if the Cauchy-Jacobi problem is solvable in $\Pi_{mn\zeta}$ for $A = A(m+n+1-d)$, $f \in M(A)$ with $d = d(f,A)$, and $\zeta \in \overline{\mathbb{C}}\backslash S(A)$, then the multipoint Padé approximant $R_{mn}(f,A,\zeta;z)$ is the solution.

The uniqueness parts of Theorem 8.3 and Lemma 8.4 show that the poles of the rational interpolant at points of $S(A)$ can be considered as prescribed poles. They are necessary to interpolate, but they do not really contribute to improve the order of contact between the function f and its interpolant.

In the definition of rational interpolants with prescribed poles for functions $f \in M(A)$ $(d(f,A) > 0)$ we assume that the poles of f on $S(A)$, which are described by $D(f,A)$, belong to the set B. We remark that therefore in contrast to the original Definition 7.2 the set B can no longer be specified independent of the function $f \in M(A)$. We describe the situation in the following lemma, which follows immediately from Lemma 7.1 and Definition 8.2.

LEMMA 8.5 *Let* $A = A(n+1) := \{a_1,\dots,a_{n+1}\}$ *and* $B = B(n) := \{b_1,\dots,b_n\}$, $n \in \mathbb{N}$, *be two sets of* $n+1$ *and* n *points* $a_j, b_j \in \overline{\mathbb{C}}$, *but unlike the assumption in Lemma 7.1 the two sets now have not to be disjoint. Let* $f \in M(A)$. *If* $D := D(f,A) \subset B$ *and* $B\backslash D$ *disjoint from* A, *then there uniquely exists a rational function* $R_n(z) = R_n(f,A,B;z) \in \Pi_{nn}$ *interpolating* f *in the set* A *and having all its poles in* B.

In summarizing the results we can say that it is possible to extend the definitions of Cauchy-Jacobi interpolants, multipoint Padé approximants, and rational interpolants with prescribed poles to interpolation of functions which assume the value infinity. In order to be able to define an order of contact it was necessary to confine ourselves to functions with only polar singularities. The poles of the interpolant that match with poles of the function f at interpolation points can be considered as being prescribed. They do not contribute to the actual order of contact between the interpolant and the function f to be interpolated.

We close this section (and the paper) by a lemma in which the existence is established of rational approximants that are intermediate between multipoint Padé approximants and rational interpolants with prescribed poles. They can be called rational interpolants with partially prescribed poles. A proof of the lemma is practically a combination of the proofs of Lemma 5.1 and 7.1.

LEMMA 8.6 Let $A = A(k)$ and $B = B(\ell)$, $k, \ell \in \mathbb{N}$, be two sets of k and ℓ points of $\overline{\mathbb{C}}$, and let $f \in M(A)$, $D := D(f,A)$, $d := d(f,A)$, and $\zeta \in \overline{\mathbb{C}} \setminus S(A)$. If $m+n = k+\ell-1$, $\ell \le n$, $D \subseteq B$, and $B \setminus D$ is disjoint from A, then there uniquely exists a rational function

$$(8.6) \qquad R_{mn}(z) = R_{mn}(f,A,B,\zeta;z) = \frac{P_{mn}(z)}{Q_{mn}(z)} \in \Pi_{mn\zeta}$$

such that $P_{mn} \in \Pi_{m\zeta}$, $Q_{mn} \in P_{n\zeta}$, R_{mn} has only $n-\ell$ poles outside of B, taking account of multiplicities, and

$$(8.7) \qquad Q_{mn}(z)f(z)-P_{mn}(z) = P_{\zeta}(A;z)h(z),$$

where h is defined and bounded in a neighborhood of $S(A)$. If $m = n$, then R_{nn} is independent of $\zeta \in \overline{\mathbb{C}} \setminus S(B)$.

References

1. Baker, G.A. Jr. and Graves-Morris, P.R. (1981): Padé Approximants, Part I and II. Encycl. of Math. Vol. 13 and 14, Cambridge Univ. Press, Cambridge.

2. Brezinski, C. (1981): The long history of continued fractions and Padé approximants. In: Padé Approximants and Applications. (de Bruin, M.G. and van Rossum, H., eds), Lect. Notes Math., Vol. 888, Springer-Verlag, Berlin 1-27.

3. Claessens, G. (1978): On the structure of the Newton-Padé table. J. Approx. Theory 22, 304-319.

4. Cauchy, A.L. (1821): Sur la formulae de Lagrange relative à l'interpolation. Analyse algebraique, Paris.

5. Jacobi, C.G.I. (1846): Ueber die Darstellung einer Reihe gegebener Werte durch eine gebrochene rationale Funktion. Crelle's J. reine u. angew. Math. 30, 127-156.

6. Kronecker, L. (1881): Zur Theorie der Elimination einer Variabeln aus zwei algebraischen Gleichungen. Monatsb. koenigl. Preuss. Akad. Wiss. Berlin, 535-600.

7. Meinguet, J. (1970): On the solubility of the Cauchy interpolation problem. In: Approximation Theory (Talbot, A., ed.), Academic Press, London, 137-163.

8. Perron, O. (1929): Die Lehre von den Kettenbruechen. Chelsea Pub. Co., New York.

ZERO-FREE DISKS IN FAMILIES OF ANALYTIC FUNCTIONS

J. Waldvogel
Applied Mathematics
ETH-Zentrum
CH-8092 Zürich
Switzerland

Abstract This note is concerned with families of functions $f(z,q)$, analytic in both the variable $z\in\mathbb{C}$ and the parameter $q\in\mathbb{C}$. In the one-parameter subfamily characterized by $q=e^{i\vartheta}$, $\vartheta\in\mathbb{R}$ we consider the problem of finding the zero $z=z_m$ of $f(z,e^{i\vartheta_m})$ with minimum modulus. An algorithm for calculating ϑ_m, z_m based on Newton's method will be described. The problem arises in several fields of approximation theory, notably in connection with certain Padé approximants and with the Whittaker and the power series constants. Conjectured values of these constants will be given in extended precision.

1. Zeros of Minimum Modulus

Certain constants in complex analysis are defined by means of zeros of analytic functions depending on a parameter. Examples are the Whittaker constant and the power series constant which are related to zeros of minimum modulus in a one-parameter family of analytic functions.

Let $f(z,\vartheta)$ be holomorphic in z in the disk $|z|<a$ and real-analytic in ϑ for $\vartheta\in\mathbb{R}$. Throughout the paper z is the principal complex variable, whereas ϑ and $q=e^{i\vartheta}$ are parameters. We consider a zero $z=Z(\vartheta)$ of $f(z,\vartheta)$ as a function of ϑ defined by

(1) $$f(Z(\vartheta),\vartheta)=0.$$

If a regular point of the curve $z=Z(\vartheta)$, i.e. a point with a nonzero tangent vector $dZ/d\vartheta \neq 0$, is at minimum distance from the origin we necessarily have

$$\frac{d}{d\vartheta}|Z(\vartheta)|=0, \quad \frac{dZ}{d\vartheta}\neq 0. \tag{2}$$

This is equivalent with the orthogonality of the radius vector $z \neq 0$ and the tangent vector, i.e.

$$\mathrm{Re}\left(\frac{1}{z}\cdot\frac{dZ}{d\vartheta}\right)=0. \tag{3}$$

By substituting $dZ/d\vartheta$ from Equ.(1), we obtain

$$\mathrm{Re}\{z^{-1}f_\vartheta(z,\vartheta)/f_z(z,\vartheta)\}=0.$$

This is, together with Equ.(1), a necessary condition for $z=Z(\vartheta)$ being a zero of $f(z,\vartheta)$ with minimum modulus:

$$f(z,\vartheta)=0, \quad \mathrm{Re}\{g(z,\vartheta)\}=0, \tag{4}$$

where

$$g(z)=\frac{f_\vartheta(z,\vartheta)}{zf_z(z,\vartheta)}, \quad z\neq 0, \quad f_z(z,\vartheta)\neq 0. \tag{5}$$

The equations (4) constitute a system of a complex and a real condition for the unknowns $z\in\mathbb{C}$, $\vartheta\in\mathbb{R}$.

Newton's method for solving the system (4) may be applied as follows. Let z_o, ϑ_o be an initial approximation sufficiently close to the solution z_m, ϑ_m. We determine increments $\Delta z\in\mathbb{C}$, $\Delta\vartheta\in\mathbb{R}$ to satisfy the linearized form of (4),

$$f(z_o+\Delta z,\vartheta_o+\Delta\vartheta)\approx f(z_o,\vartheta_o)+\Delta z\cdot f_z+\Delta\vartheta\cdot f_\vartheta=0$$

$$\mathrm{Re}\{g(z_o+\Delta z,\vartheta_o+\Delta\vartheta)\}\approx\mathrm{Re}\{g(z_o,\vartheta_o)+\Delta z\cdot g_z+\Delta\vartheta\cdot g_\vartheta\}=0.$$

Omitting the arguments z_o, ϑ_o, we obtain from the first equation

$$\Delta z=-\frac{f+\Delta\vartheta\cdot f_\vartheta}{f_z}. \tag{6}$$

Substituting this into the second condition yields

$$(7) \qquad \Delta\vartheta = -\frac{\mathrm{Re}(g - af)}{\mathrm{Re}(g_\vartheta - af_\vartheta)}$$

where

$$(8) \qquad a = g_z/f_z;$$

Δz may be calculated from Equ.(6). In the usual way we define $z_{j+1}=z_j+\Delta z$, $\vartheta_{j+1}=\vartheta_j+\Delta\vartheta$, $(j=0,1,\ldots)$; then $\lim_{j\to\infty} z_j=z_m$, $\lim_{j\to\infty} \vartheta_j=\vartheta_m$. Generally, the convergence rate is quadratic if the initial guess is sufficiently close to the solution.

2. The Whittaker Constant

The Whittaker constant [4, 5], W, is the least exponential type of an entire function, not identically zero, each of whose derivatives has a zero in the closed unit disk. Following Macintyre [8], an upper bound $W_o \geq W$ may be obtained by constructing an analytic function $\mu(z,q)$ satisfying the functional differential equation

$$(9) \qquad \mu'(z,q)=\mu(qz,q), \qquad ' \equiv \frac{d}{dz}$$

with a parameter $q \in \mathbb{C}$. A series solution is given by the Macintyre function

$$(10) \qquad \mu(z,q) = \sum_{n=0}^{\infty} c_n q^{\binom{n}{2}} z^n, \qquad c_n = \frac{1}{n!};$$

it is an entire function in z for $|q| \leq 1$. Repeated application of (9) yields

$$(11) \qquad \mu^{(j)}(zq^{-j},q)=\mu(z,q).$$

If z_q is a zero of $\mu(z,q)$ we have

$$\mu^{(j)}(z_q q^{-j},q)=0, \qquad j=0,1,\ldots \quad .$$

For the zeros of $\mu^{(j)}(z,q)$ to lie in a bounded region we assume $|q|=1$, $q=e^{i\vartheta}$; then every derivative of $\mu(z,q)$ with respect to z has a zero in the disk $|z|\le|z_q|$ (in fact on the circle $|z|=|z_q|$). The substitution $z=\zeta|z_q|$ shows that every derivative of $\mu(\zeta|z_q|,q)$ has a zero in $|\zeta|\le 1$. Since $\mu(\zeta|z_q|,q)$ has the exponential type $|z_q|$, the least upper bound W_o for W is given by

$$W_o=\inf_{\vartheta}|z_q|, \quad \mu(z_q,e^{i\vartheta})=0. \tag{12}$$

For calculating W_o we use the methods of Section 1 with $f(z,\vartheta)=\mu(z,\exp(i\vartheta))$. First, algorithmic aspects of the efficient evaluation of $f(z,\vartheta)$, $g(z,\vartheta)$ are discussed. With

$$z=re^{i\varphi}, \quad q=e^{i\vartheta} \tag{13}$$

we rewrite Equ.(10) as

$$f(z,\vartheta)=\mu(re^{i\vartheta},q) = \sum_{n=0}^{\infty} p_n\cdot e_n, \tag{14}$$

where

$$p_n=c_n r^n, \quad e_n=q^{\binom{n}{2}} e^{in\varphi}.$$

The auxiliary quantities p_n, e_n may be calculated recursively by

$$\begin{aligned} &p_o=1, && p_n=\frac{r}{n}p_{n-1} \\ &d_o=e^{i\varphi}, && d_n=d_{n-1}\cdot q \qquad n=1,2,\ldots \\ &e_o=1, && e_n=e_{n-1}\cdot d_{n-1} \end{aligned} \tag{15}$$

with only 2 complex multiplications per cycle. Equ.(14) and the relations

$$\begin{aligned} zf_z &= \sum_{n=0}^{\infty} np_n e_n \\ f_\vartheta &= \frac{i}{2}\sum_{n=0}^{\infty} n(n-1)p_n e_n \end{aligned} \tag{16}$$

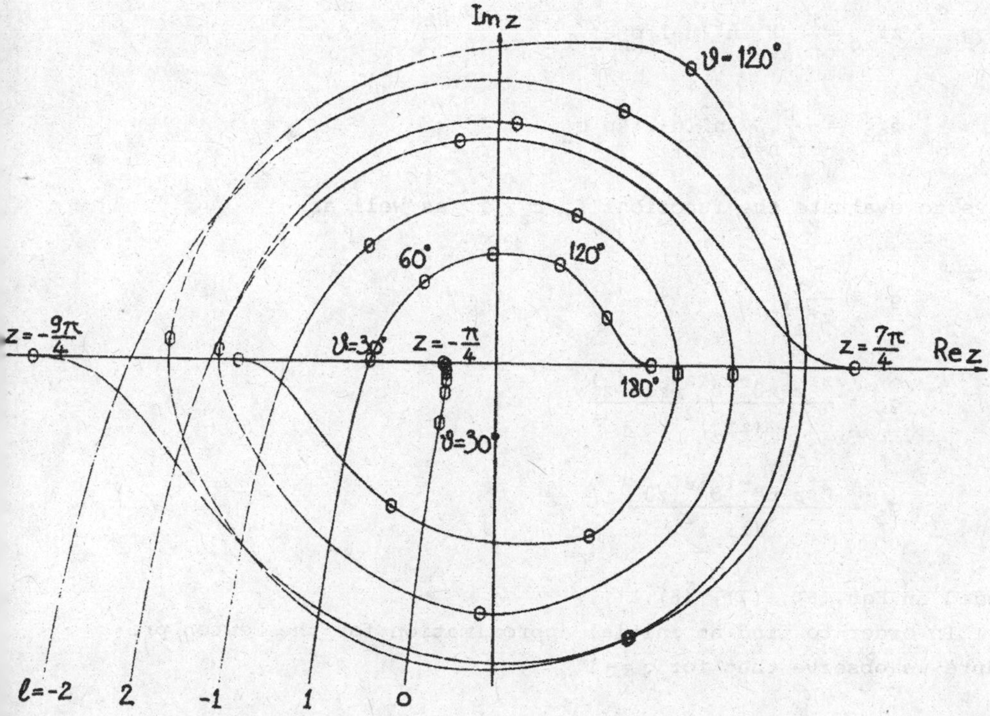

Figure 1

Locus $z=Z(\vartheta)$ of zeros of the Macintyre function $\mu(z,e^{i\vartheta})$, Branches $\ell=0, \pm 1, \pm 2$. The marked points correspond to $\vartheta = k\frac{\pi}{6}$ $(k=1,2,\ldots,6)$.

$$zf_{z\vartheta} = \frac{i}{2} \sum_{n=0}^{\infty} n^2(n-1)p_n e_n$$

(16)

$$f_{\vartheta\vartheta} = -\frac{1}{4} \sum_{n=0}^{\infty} n^2(n-1)^2 p_n e_n$$

serve to evaluate the functions f, f_z, f_ϑ as well as

$$g = \frac{f_\vartheta}{zf_z},$$

(17)
$$g_z = \frac{zf_z f_{z\vartheta} - f_\vartheta(zf_{zz} + f_z)}{(zf_z)^2}$$

$$g_\vartheta = \frac{zf_z f_{\vartheta\vartheta} - f_\vartheta \cdot zf_{z\vartheta}}{(zf_z)^2}$$

needed in Equ. (6), (7), (8).

In order to find an initial approximation for the Newton procedure we observe that for $q = -1$

(18) $$\mu(z,-1) = \cos z + \sin z$$

is an elementary function; its zeros $z = \zeta_\ell$ are given by

(19) $$\zeta_\ell = (\ell - \tfrac{1}{4})\pi, \quad \ell = 0, \pm 1, \pm 2, \ldots \; .$$

For different values of ℓ the points $z = \zeta_\ell$ belong to different branches of the locus $z = Z(\vartheta)$, shown in Fig.1 for $\ell = 0, \pm 1, \pm 2$. The points $z = \zeta_\ell$ are cusps on their respective branches with $dZ/d\vartheta = 0$, and it may be shown that for ϑ near π asymptotic (divergent) expansions

(20) $$Z(\vartheta) = \zeta_\ell + \frac{1}{2}\zeta_\ell^3 u^2 + \frac{1}{6}\zeta_\ell^3 u^3 - \frac{7}{24}\zeta_\ell^5 u^4 - \frac{27}{20}\zeta_\ell^5 u^5 + O(u^6)$$

with

$$u = -e^{i\vartheta} - 1$$

```
R,PHI,THETA,MAC = .6   3.25   2.4   1
  1 TH NEWTON STEP: F1,F2,G1,R,PHI,THETA
   0.232147243137096442    -0.004362803350251495    -0.017581216715036596
   0.745210684477802324     3.256538707122259072     2.385790400498541999
  2 TH NEWTON STEP: F1,F2,G1,R,PHI,THETA
  -0.013256427279549484     0.000489593725470152    -0.004699918441401878
   0.737766270265686565     3.254515312544278977     2.393143648502567212
  3 TH NEWTON STEP: F1,F2,G1,R,PHI,THETA
  -0.000026623967510066     0.000028771921507925    -0.000083252076334316
   0.737750755617329281     3.254509368576922114     2.393242276113060276
  4 TH NEWTON STEP: F1,F2,G1,R,PHI,THETA
   0.000000003360869378     0.000000001795913231     0.000000000442201325
   0.737750757478224333     3.254509370576366159     2.393242274134896564
  5 TH NEWTON STEP: F1,F2,G1,R,PHI,THETA
  -0.000000000000000001    -0.000000000000000006    -0.000000000000000004
   0.737750757478224333     3.254509370576366152     2.393242274134896572
 23 TERMS IN MACSAF

R,PHI,THETA,MAC = .6   3.25   2.4   0
  1 TH NEWTON STEP: F1,F2,G1,R,PHI,THETA
  -0.126627563255909201    -0.015744224854526055     0.461952196653292372
   0.560756868888504938     3.273213452819635879     2.373180189913288108
  2 TH NEWTON STEP: F1,F2,G1,R,PHI,THETA
   0.001407614406193649    -0.008802208852867649    -0.004277026315106289
   0.561214943599224829     3.266659367911767547     2.375160735148353996
  3 TH NEWTON STEP: F1,F2,G1,R,PHI,THETA
   0.000044696622981742     0.000008414579040442     0.000413971621896432
   0.561228761211093252     3.266694748822403566     2.375125269985822156
  4 TH NEWTON STEP: F1,F2,G1,R,PHI,THETA
   0.000000005635064212    -0.000000000614071967     0.000000194203405939
   0.561228762962881702     3.266694762331483245     2.375125253834255203
  5 TH NEWTON STEP: F1,F2,G1,R,PHI,THETA
   0.000000000000001414    -0.000000000000000459     0.000000000000053135
   0.561228762962882143     3.266694762331486738     2.375125253834250831
101 TERMS IN MACSAF
```

Table 1

5 Newton steps in 19D accuracy with the initial values $r=0$, $\varphi=3.25$, $\vartheta=2.4$. On the first line in each block the residuals f, Re(g) are listed. MAC=1: Macintyre function, MAC=0: partial theta function.

```
   0 TH NEWTON STEP: RR, PHI, THETA
 0.73775075746666666666666666666666666666666666666666666666666666666
 3.25450937060000000000000000000000000000000000000000000000000000000
 2.39324227413333333333333333333333333333333333333333333333333333333
   87 TERMS IN MACSAF. RESIDUALS: F1, F2, G1
 0.00000000001953427851853717914146358791580302311975650295156420219
-0.00000000003101012347073132147153012832950173268732421888660233035
 0.00000000001005404822147898288204977407970137012278884910540214218
   1 TH NEWTON STEP: RR, PHI, THETA
 0.73775075747822433247306574232809914708115304877085991993980725674
 3.25450937057636615233810639274183522732577947253875400596805958672
 2.39324227413489657259188255437340186302830089167533365181004137532
   87 TERMS IN MACSAF. RESIDUALS: F1, F2, G1
 0.00000000000000000000047035812251556366814999803620642256017286877
 0.00000000000000000000071334379822164071726562444844655129747214213
-0.00000000000000000000021194725664685140096174823256608809736931324
   2 TH NEWTON STEP: RR, PHI, THETA
 0.73775075747822433247331791037337232641749612391141046596232554530
 3.25450937057636615233870862941070506388014628453683454357017442955
 2.39324227413489657259171538828124240601345804694573459132520455052
   87 TERMS IN MACSAF. RESIDUALS: F1, F2, G1
 0.00000000000000000000000000000000000000000023004704827203388974182
-0.00000000000000000000000000000000000000000037878567585256559066294
 0.00000000000000000000000000000000000000000003664184998095923461864
   3 TH NEWTON STEP: RR, PHI, THETA
 0.73775075747822433247331791037337232641749626026442902301521425051
   89794141855760656206756262033884032190444031024233394601346647708
 3.25450937057636615233870862941070506388014596344383741179515107794
   91555093435474675406623597530082465769885672143102888245746475485
 2.39324227413489657259171538828124240601345818253308566172258942443
   45748435823443165092563777716061036181716025742990488322430462744
   87 TERMS IN MACSAF. RESIDUALS: F1, F2, G1
 0.00000000000000000000000000000000000000000000000000000000000000000
   00000000000000000000646565744948381610199502778398464940801281563
 0.00000000000000000000000000000000000000000000000000000000000000000
   00000000000000000000934205442454666005518730201208052251248812178
-0.00000000000000000000000000000000000000000000000000000000000000000
   0000000000000000000064303160445061993279071885636642495515082900
   4 TH NEWTON STEP: RR, PHI, THETA
 0.73775075747822433247331791037337232641749626026442902301521425051
   89794141855760656207103728159772775213908097446512043192576833647
 3.25450937057636615233870862941070506388014596344383741179515107794
   91555093435474675407493030145190336505808428826853425838723360379
 2.39324227413489657259171538828124240601345818253308566172258942443
   45748435823443165092055458595643765310923708482243856473503629473
```

Table 2

Macintyre function: 4 Newton steps in 130D accuracy with 10D initial values using Brent's MP package. Lines 6 and 5 from the bottom show the conjectured value of the Whittaker constant in full precision.

exist in every cusp. The moduli $|Z(\vartheta)|$ at the cusps are local <u>maxima</u>.

The point $z_m=Z(\vartheta_m)$ closest to the origin is on the branch $\ell=0$. Since it is a regular point with $dZ/d\vartheta \neq 0$ the Newton procedure described above may be used to precisely locate ϑ_m and $z_m=r_m\cdot\exp(i\varphi_m)$ with the initial approximation

(21) $r_o=0.6, \quad \varphi_o=3.25, \quad \vartheta_o=2.4.$

4 Newton steps yield an accuracy of 16 decimals (see Table 1):

$$r_m=0.73775\ 07574\ 78224$$
$$\varphi_m=3.25450\ 93705\ 76366$$
$$\vartheta_m=2.39324\ 22741\ 34896$$

By beginning with a 10 digit initial approximation more than 150 decimals may be obtained with 4 Newton steps. By means of R. Brent's multiple precision arithmetic [1] 130 decimals were obtained within 55 seconds of computer time (see Table 2). According to a conjecture by Varga and Waldvogel (the VW conjecture, see [10], p.124, 136) $W_o=r_m$ is the actual Whittaker constant.

3. The Power Series Constant

The problem of the power series constant [2, 3] bears much similarity with the subject of the Whittaker constant, although the analytic function involved is more complicated.

Let

$$s_n(z) = \sum_{k=0}^{n} a_k z^k$$

be the n-th partial sum of the series

$$F(z) = \sum_{k=0}^{n} a_k z^k,$$

and let

$$r_n(z) := z^{-n}[F(z)-s_{n-1}(z)] = a_n + a_{n+1}z + \ldots$$

be the n-th normalized remainder of $F(z)$; $r_o(z)=F(z)$. We consider the class $\mathcal{F}$ of functions $F(z)$, analytic in $|z|<\rho, \rho>1$, not identically zero, with the property that for every $n\geq 0$ the n-th normalized remainder has a zero in the closed unit disk. The power series constant P [6] is defined as

$$P = \sup_{F\in\mathcal{F}} \rho,$$

i.e. as the largest possible radius of convergence in the class $\mathcal{F}$.

Hence the role of differentiation in the Whittaker problem is taken over by the shift operator $\mathcal{S}$ defined by

$$\mathcal{S}r_n(z) = r_{n+1}(z), \quad n=0,1,\ldots .$$

In analogy to (9) we consider the functional equation

$$(22) \qquad (\mathcal{S}M)(z,q) = M(qz,q)$$

for the function $M(z,q)$, $q\in\mathbb{C}$ being a complex parameter. Equ.(22) is solved by the series

$$(23) \qquad M(z,q) = \sum_{n=0}^{\infty} q^{\binom{n}{2}} z^n, \quad |q|\leq 1,$$

referred to as the partial theta function. As in Section 2 we consider the case $|q|=1$, $q=e^{i\vartheta}$ in order to keep the zeros of the normalized remainders on a circle. Analoguous considerations show

$$(24) \qquad P^{-1} \leq \inf_{q=e^{i\vartheta}} |z_q|, \quad M(z_q,q)=0.$$

The calculation of the least upper bound of P^{-1} thus again reduces to the problem of Section 1, the computation of the zero of minimum modulus in the family $M(z,e^{i\vartheta})$. Again, it is conjectured that P^{-1} equals the least upper bound.

The partial theta function $M(z,e^{i\vartheta})$ and its derivatives may be evaluated according to Equ.(10) with $c_n=1$. In the algorithm (14),

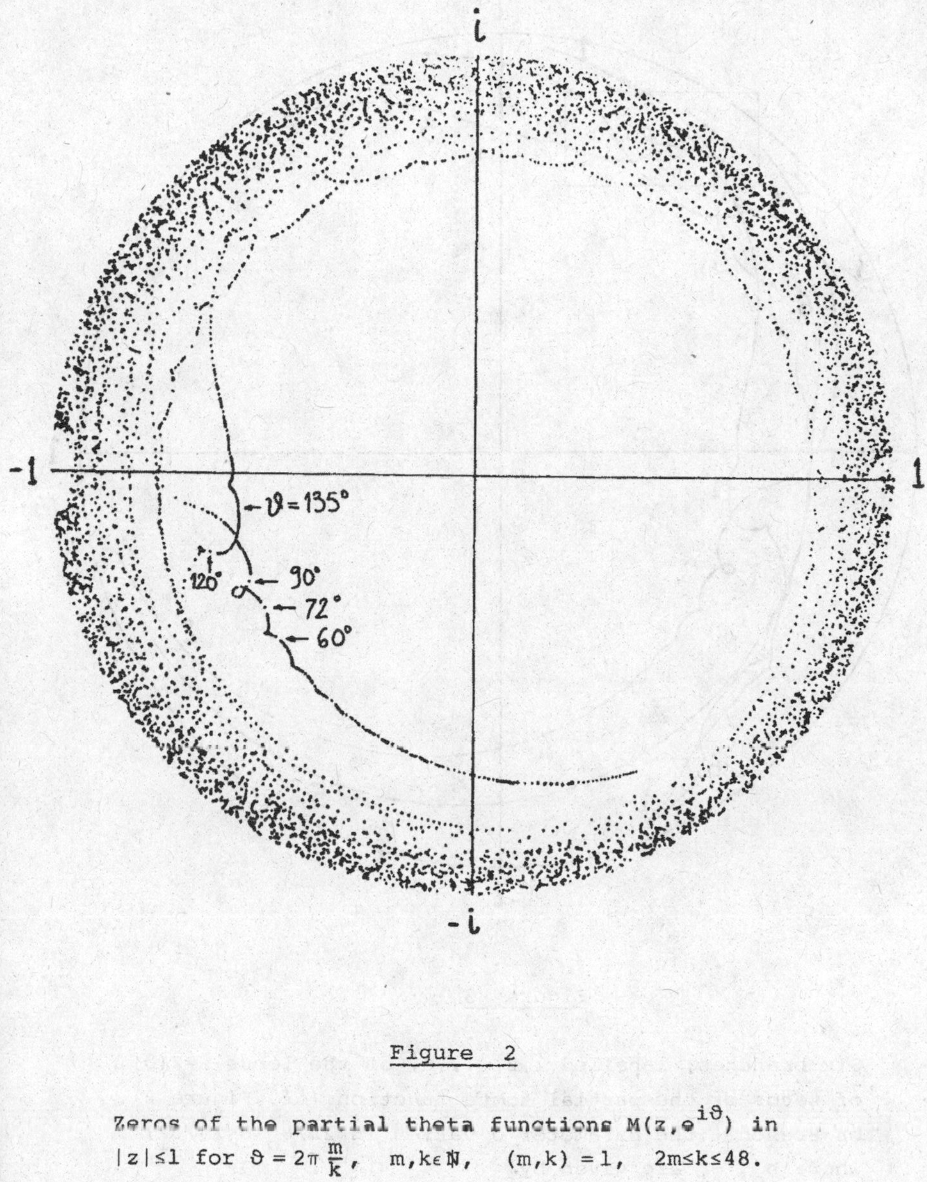

Figure 2

Zeros of the partial theta functions $M(z,e^{i\vartheta})$ in $|z|\le 1$ for $\vartheta = 2\pi\frac{m}{k}$, $m,k\in\mathbb{N}$, $(m,k)=1$, $2m\le k\le 48$.

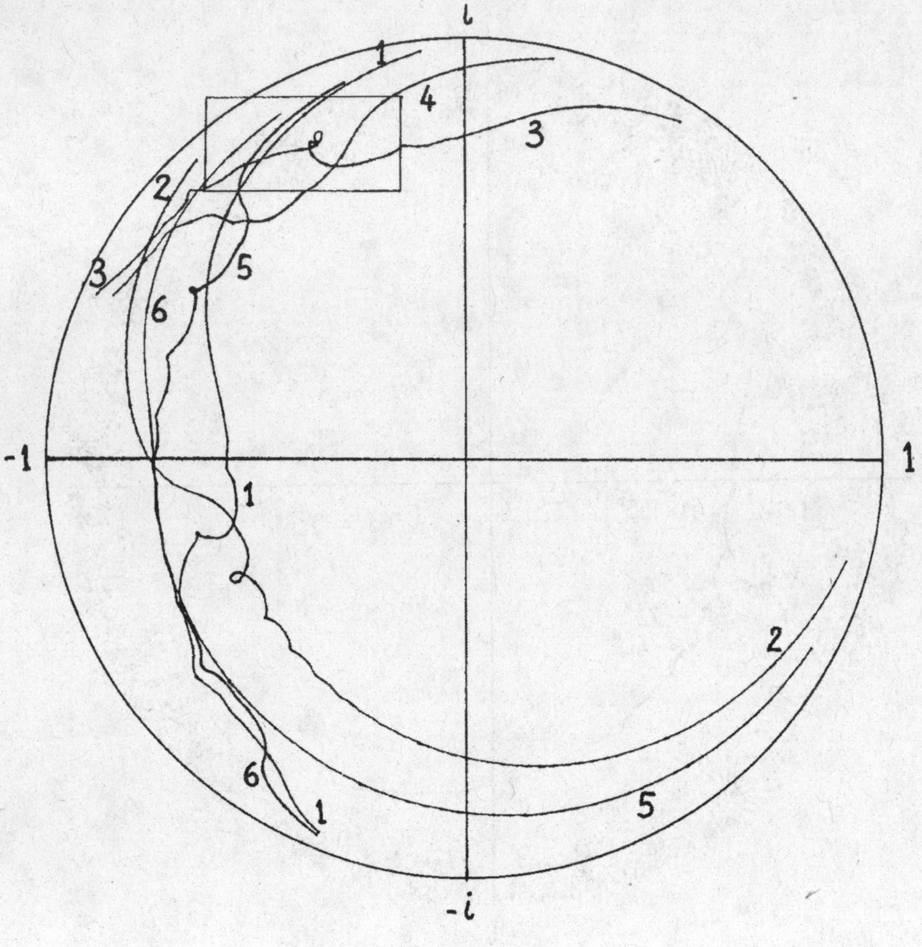

Figure 3

Six branches, labelled 1,2,...,6, of the locus $z=Z(\vartheta)$ of zeros of the partial theta function (cf. Figure 2). On Branch ℓ the parameter ϑ varies in $2\pi/\rho_o<\vartheta<2\pi/\rho_1$, where ρ_o, ρ_1 arc given by

ℓ	1	2	3	4	5	6
ρ_o	3.5	∞	5.5	6.5	∞	2.333..
ρ_1	2	3	4	5	6	2

(15), (16) the only modification is $p_n = rp_{n-1}$ instead of the second equation (15). Due to the slow convergence, however, more terms are needed than in the Macintyre case (see Tables 1, 2, 3).

In order to discuss the locus $Z(\vartheta)$ of zeros of $M(z,q)$ we observe that if q is an N-th root of unity, $q^N=1$, $M(z,q)$ is the rational function

$$(25)\qquad M(z,q) = \frac{\sum_{n=0}^{N-1} q^{\binom{n}{2}} z^n}{1+(-z)^N}, \quad q^N=1.$$

Therefore every point of the unit circle is an accumulation point of poles; $|z|=1$ is a natural boundary. The locus $Z(\vartheta)$ splits up into infinitely many branches that are connected at points on the unit circle. A rough picture may be obtained by plotting the roots of $M(z,q)$ for some rational cases (25). In Fig.2 all zeros in $|z|\leq 1$ are plotted (by means of a polynomial root finder) if q runs through the primitive N-th ($N\leq 48$) roots of unity in the upper half plane,

$$q=e^{2\pi i\frac{m}{k}}, \quad m,k\in\mathbb{N}, \quad (m,k)=1, \quad 2m\leq k\leq 48.$$

An amazing complexity is observed; in Fig.3 six of the infinitely many branches are traced by varying the parameter ϑ continuously; Fig.4 shows a detail on branch 3.

The point $z_m=Z(\vartheta_m)$ of minimum modulus $r_m=|z_m|$ is on branch 1; it may be computed by the Newton procedure using the initial values (21). After 4 steps an accuracy of 14 decimals is achieved (Table 1, MAC=0). The results of a multiple precision computation in 65D arithmetic are shown in Table 3: 10D initial values yield full precision in 3 Newton steps (47 seconds). The conjectured power series constant $P=r_m^{-1}$ is found to be

P=1.7818046151 4610000210 2441189850 8399163778
3536137062 8651392041 5791044543 3036

Finally, in Table 4 some further minima and maxima of $|Z(\vartheta)|$ on the branches 1, 2 and 3 are given.

```
    0 TH NEWTON STEP: RR, PHI, THETA
 0.56122876296666666666666666666666666666666666666666666666666666667
 3.26669476233333333333333333333333333333333333333333333333333333333
 2.37512525383333333333333333333333333333333333333333333333333333333
    310 TERMS IN MACSAF. RESIDUALS: F1, F2, G1
-0.00000000012228156291426693322853308918108235267513603776395292l0
-0.0000000000170387999003528021856882700229898819098762818587566515
-0.0000000000101295970675119606413284525964469208032769372202957024 7
    1 TH NEWTON STEP: RR, PHI, THETA
 0.56122876296288214325407817063548464298259377378441665484639068330
 3.26669476233148673777754643058803694071787955154034662001416232555
 2.37512525383425083176258635825513292695102901075802282126532349853
    310 TERMS IN MACSAF. RESIDUALS: F1, F2, G1
-0.0000000000000000000006384757790715341497704973179142895742861904
-0.0000000000000000000001472796082675657542648878289201722147737874
-0.0000000000000000000004356902736799258842848662130866489296535511 2
    2 TH NEWTON STEP: RR, PHI, THETA
 0.56122876296288214325405844270326190149807634516990181592170556674
 3.26669476233148673777750515569793816989978972640219608284213572075
 2.37512525383425083176262441674287245356776529762207169258810203342
    310 TERMS IN MACSAF. RESIDUALS: F1, F2, G1
 0.0000000000000000000000000000000000000000000005497818570872945189 88
-0.0000000000000000000000000000000000000000000003320203499823491804 3
 0.0000000000000000000000000000000000000000000001914864040109035109 2805
    3 TH NEWTON STEP: RR, PHI, THETA
 0.56122876296288214325405844270326190149807634687754033209316443691
 3.26669476233148673777750515569793816989978973991431261488297527376
 2.37512525383425083176262441674287245356776528164795085331032464461
```

<u>Table 3</u>

Partial theta function: 3 Newton steps in 65D accuracy with 10D initial values using Brent's MP package.

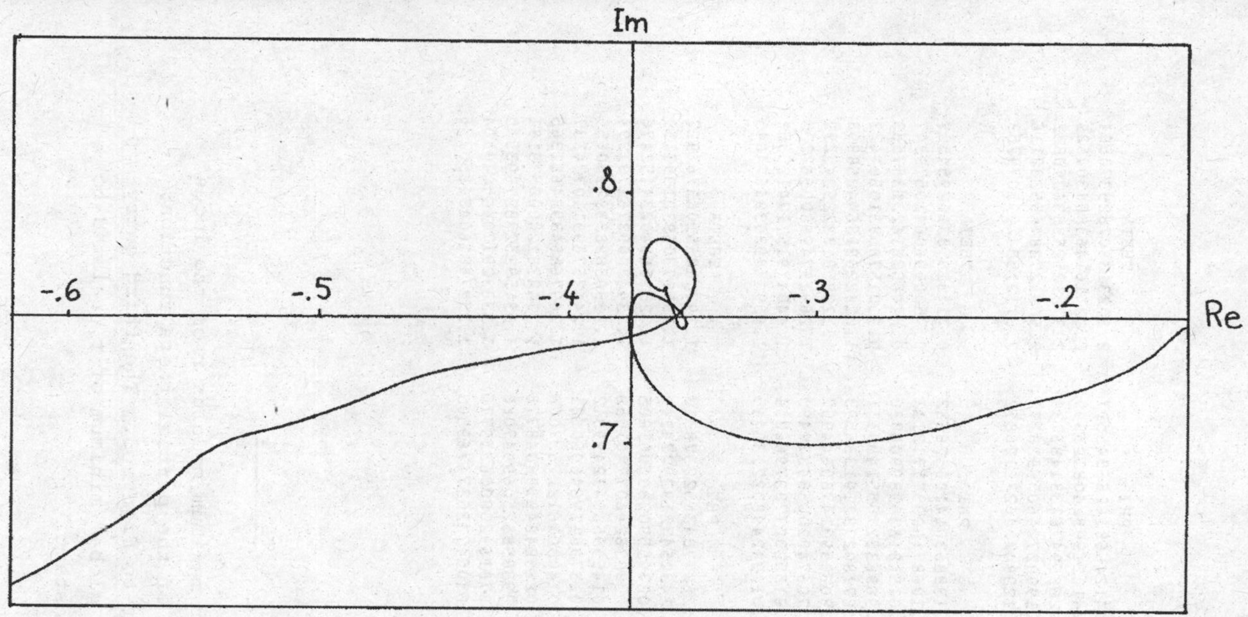

Figure 4

Detail of Branch 3 (cf. Figure 3), magnified 6 times. See also Table 4, where 11 points of maximum or minimum modulus on Branch 3 are listed (2 to the left, 6 inside, and 3 to the right of the loop structure).

BRANCH 1: R	PHI	THETA
0.667906069138675797	3.412466836266545597	2.002313807829370594
0.671743535555103052	3.412219052408125999	2.021161463884304529
0.561228762962882143	3.266694762331486738	2.375125253834250832
0.578338743431621781	3.190027246056928173	2.509127884655388116
0.568570990517021454	3.129842246272126872	2.611215342998079753

BRANCH 2: R	PHI	THETA
0.635279289469427946	4.138863014108526632	0.717329583102954507
0.636694979161729335	4.106871099571712111	0.750099606410579347
0.620066800287038269	4.015181829989906836	0.844653616231387563
0.625434088495967258	3.969633275551499111	0.900157858310587473
0.603459008883140924	3.871802732191336593	1.019220898644258893
0.620764614790756410	3.806099314780644957	1.122058153229245292
0.586171301357213410	3.706748200684874490	1.269197499800567235
0.637179830969838162	3.615520277827966034	1.440479167174195358
0.570744512295991208	3.512942498022132445	1.659978930394674645

BRANCH 3: R	PHI	THETA
0.882505907398299988	2.262378436520578313	1.192788651063762917
0.883338949052825806	2.243254721939274144	1.195193069301398159
0.829882965453703420	2.024955902806001805	1.225706940234152476
0.861811653889756095	2.008836246077401052	1.234480892761791293
0.842061840020202304	2.014599580432499312	1.239088883397900655
0.843058313610124430	2.013798129646660631	1.241357699110774337
0.824862963799068952	2.014030161217537008	1.269765543984681365
0.845110287776979659	2.026584849198048778	1.274238120005579181
0.748041661260719478	1.893858525069947088	1.331581527827093530
0.748225526254975665	1.878169208006359716	1.338328170460040104
0.746768177355956993	1.840547420374769503	1.354748160486850736

Table 4

Points of minimum or maximum modulus r on the locus $r e^{i\varphi}=Z(\vartheta)$ of zeros of the partial theta function. Branch numbers refer to Figure 3. The third entry for Branch 1 is the global minimum of r, related to the power series constant.

4. The Rogers-Szegö Polynomials

In [7] Lubinsky and Saff show that the denominators $Q_{mn}(z)$ of the Padé approximants of the partial theta function $M(z,e^{i\vartheta})$ are the Rogers-Szegö polynomials $G_n(\zeta)$, evaluated at $\zeta=-zq^m$ (if $m\geq n-1\geq 0$, $q^j \neq 1 \;\forall j=1,2,\ldots,n$). Therefore the smallest zeros of $G_n(z)$ are intimately connected with the smallest poles of these Padé approximants. Since the algorithm described above readily produces the smallest zeros of $G_n(z)$ we list some of them in Table 5.

The Rogers-Szegö polynomials [9] depend on the parameter $q=e^{i\vartheta}$ which is not displayed explicitly in the notation $G_n(z)$. For q,z fixed, the sequence $G_n(z)$ may be defined recursively by $G_{-1}(z)=0$, $G_o(z)=1$ and

(26) $$G_{n+1}(z)=(1+z)G_n(z)-(1-q^n)zG_{n-1}(z), \quad n=0,1,\ldots .$$

Hence

(27) $$G_1(z)=1+z, \quad G_2(z)=1+z(1+q)+z^2, \quad G_3(z)=(1+z)[1+z(q+q^2)+z^2].$$

The Rogers-Szegö polynomials satisfy the functional equation

(28) $$G_{n+1}(z)=G_n(qz)+zG_n(z), \quad G_n(0)=1, \quad n=0,1,\ldots .$$

By introducing

(29) $$\Delta_n=G_{n+1}-G_n, \quad A_n = \begin{pmatrix} 1 & 1 \\ q^n z & z \end{pmatrix},$$

Equ.(26) may be conveniently written as

(30) $$\begin{pmatrix} G_n \\ \Delta_n \end{pmatrix} = A_n \begin{pmatrix} G_{n-1} \\ \Delta_{n-1} \end{pmatrix}, \quad n=0,1,\ldots , \quad \begin{pmatrix} G_{-1} \\ \Delta_{-1} \end{pmatrix} = \begin{pmatrix} 0 \\ 1 \end{pmatrix}.$$

For the derivatives with respect to z and q similar relations may be derived by differentiation, e.g.

(31) $$\begin{pmatrix} G_{n,q} \\ \Delta_{n,q} \end{pmatrix} = A_n \begin{pmatrix} G_{n-1,q} \\ \Delta_{n-1,q} \end{pmatrix} + \begin{pmatrix} 0 \\ nq^{n-1}z\, G_{n-1} \end{pmatrix}, \quad \begin{pmatrix} G_{-1,q} \\ \Delta_{-1,q} \end{pmatrix} = 0 .$$

N	R	PHI	THETA
2	0.57735026918962576 4	2.18627603546528396 0	1.23095941734077468 2
3	0.448939424135778226	1.757621771757933056	0.862437676064527718
4	0.386203389385391771	1.498783103351402079	0.667099742598454402
5	0.348667985143413301	1.320768224240116513	0.545255814882549535
6	0.323511334541452956	1.188899098264904703	0.461669941550720635
7	0.305379897759588054	1.086350236017668482	0.400625613807163072
8	0.291634276218761031	1.003798834768468701	0.354019988314428087
9	0.280818870041677258	0.935596919208420345	0.317237247273555208
10	0.272063167193821353	0.878094957169711830	0.287448739065771076
11	0.264813556811340842	0.828816235655157487	0.262821469421355084
12	0.258700545388248502	0.786013596760678738	0.242114051549434950
15	0.244956163321505953	0.685253554687811203	0.195925616733656249
16	0.241440272742709025	0.658399190126035479	0.184234268682038873
18	0.235490385618542153	0.611877303027253266	0.164612201849958519
20	0.230635230639997571	0.572864707816813035	0.148786625115025237
24	0.223154559278136554	0.510792073037832218	0.124823166497863774
25	0.221624385908419624	0.497787652718916032	0.119996627426054842
30	0.215363324888167885	0.443429899528809009	0.100571806996471229
32	0.213353341414725835	0.425574615144010223	0.094461256490483911
36	0.209935234189935546	0.394739163777765520	0.084232490498452157
40	0.207129949688263645	0.368974807063714817	0.076008105371462793
45	0.204250114810441028	0.342081960435130443	0.067745538767236474
48	0.202777964109007328	0.328156050574619323	0.063599483297223830
50	0.201883157834401115	0.319631358880753711	0.061107008091531940
54	0.200269092832850621	0.304137464967478772	0.056666803028874318
60	0.198204838931838279	0.284098832151319374	0.051099721874144110
64	0.197018027559855244	0.272462267153731806	0.047959814183084008
72	0.194992889090701214	0.252406826145743697	0.042712762293314196
80	0.193324186062822062	0.235688734825289545	0.038502279064541393
81	0.193135788880313332	0.233790153645015433	0.038033726156006541
100	0.190186503194882815	0.203769019519371203	0.030893398515453619
200	0.183108240419598061	0.129318640721101483	0.015549678607176115
256	0.181316018023689132	0.109902125184222734	0.012168346903036276
320	0.179941400403028289	0.094847852069000007	0.009746957643436622
400	0.178763935115259568	0.081838689200281440	0.007805915828310421
500	0.177754536416216940	0.070601275873182827	0.006250410980825055
600	0.177037289814094009	0.062567931344373796	0.005212011154017649
625	0.176888610818909948	0.060897614048763794	0.005004191389252118
640	0.176804120566453273	0.059947637957252912	0.004887271815442816
750	0.176272967868870221	0.053962568358571776	0.004172428969291926
800	0.176072766776588485	0.051700847757351579	0.003912340458719118
801	0.176068977904813147	0.051658013019461119	0.003907469130604918
900	0.175729343314856724	0.047813626387464833	0.003478681511415830
999	0.175447229702943578	0.044613272911766895	0.003134717483930861
1000	0.175444620792723253	0.044583646935739898	0.003131589865810147
1001	0.175442016260926322	0.044554070131066701	0.003128468484477001
1080	0.175249159992038283	0.042362488901861043	0.002900111937277660
1296	0.174824894337962976	0.037530596333195266	0.002417644575382908

<u>Table 5</u>

Zeros $z=r\,e^{i\varphi}$ of minimum modulus for the Rogers-Szegö polynomials $G_n(z)$; $q=e^{i\vartheta}$ is the parameter value associated with the minimum.

Working upwards in a loop over n and handling the derivative equations (31), etc., before Equ.(30) yields an efficient evaluation procedure for $G_n(z)$ and its derivatives.

As a check, we observe that the smallest zero of $G_2(z)$ may be found by elementary methods. With $z=re^{i\varphi}$, $q=e^{i\vartheta}$ we obtain

$$r=\frac{1}{\sqrt{3}}\,,\quad \varphi=\arccos(-r)\,,\quad \vartheta=\arccos(\tfrac{1}{3})$$

$$z=\frac{-1+\sqrt{2}\,i}{3}\,,\quad q=\frac{1+2\sqrt{2}\,i}{3}\,.$$

In Table 5 the polar coordinates r_n, φ_n, ϑ_n of the smallest zero of $G_n(z)$ are listed for some particular values of n. It was easily generated by choosing suitable initial values to make sure that the smallest zeros are found. Fitting the data of Table 5 with asymptotic series leads to the conjecture

$$r_n=\frac{1}{3+2\sqrt{2}}+0.3833\;n^{-2/3}+O(n^{-4/3}),$$

$$\varphi_n=\qquad 4.476\;n^{-2/3}+O(n^{-4/3}),$$

$$n\,\vartheta_n=\pi-0.8576\;n^{-2/3}+O(n^{-1}),\qquad n\to\infty\,.$$

The starting point for a proof could be looking at the behaviour of $G_n(z)$ for the parameter value $q=e^{i\pi/n}$:

$$G_n(z)=\sum_{k=0}^{n}\frac{\exp[\frac{\pi i}{2}k(1-\frac{k}{n})]\;z^k}{\cos(\frac{k}{n}\frac{\pi}{2})\prod_{j=1}^{k}\tan(\frac{j}{n}\frac{\pi}{2})}\,.$$

Acknowledgements

The author wishes to thank Ed Saff for bringing up the problems discussed in Sections 3 and 4 and for the many stimulating discussions in Zürich and in Tampa. Sincere gratitude is extended to Richard Varga for mentioning the problems of Section 2 and for the numerous helpful ideas.

References

1. Brent, R.P.: A FORTRAN Multiple Precision Arithmetic Package. ACM Trans. on Math. Software 4 (1978), 57-81.

2. Buckholtz, J.D.: Zeros of Partial Sums of Power Series. Michigan Math. J. 15 (1968), 481-484.

3. Buckholtz, J.D.: Zeros of Partial Sums of Power Series II. Michigan Math. J. 17 (1970), 5-14.

4. Buckholtz, J.D., Frank, J.L.: Whittaker Constants. Proc. London Math. Soc. 23 (1971), 348-370.

5. Buckholtz, J.D., Frank, J.L.: Whittaker Constants II. Journal of Approx. Th. 10 (1974), 112-122.

6. Crofts, G.W., Shaw, J.K.: Successive Remainders of the Newton Series. Trans. Amer. Math. Soc. 181 (1973), 369-383.

7. Lubinsky, D.S., Saff, E.B.: Convergence of Padé Approximants of Partial Theta Functions and the Rogers-Szegö Polynomials, Const. Approx. 3 (1987), 331-361.

8. Macintyre, S.S.: An Upper Bound for the Whittaker Constant W. J. London Math. Soc. 22 (1947), 305-311.

9. Szegö, G.: Ein Beitrag zur Theorie der Thetafunktionen. Sitzungsberichte der Preuss. Akad. Wiss., Phys. Math. Kl. (1926), 242-251.

10. Varga, R.S.: Topics in Polynomial and Rational Interpolation and Approximation. Les Presses de l'Université de Montreal, NATO Advanced Study Institute, Montreal 1982.